Ian Lawton

P2MS Press

First published in 2015 by P2MS Press.
Second edition 2017.

All enquiries to be directed to www.prince2madesimple.co.uk.

Diagrammatic material taken from "Managing Successful Projects with PRINCE2® 2017 Edition" is reproduced under licence from AXELOS Limited, copyright © AXELOS Limited. All rights reserved.

PRINCE2® is a Registered Trade Mark of AXELOS Limited.

Copyright © Ian Lawton 2015, 2017 unless otherwise stated.
All rights reserved.
The rights of Ian Lawton to be identified as the Author of the Work have been asserted by him in accordance with the Copyright, Designs and Patents Act 1988.

No proprietary content in this book, in particular the diagrams, may be used or reproduced in any manner whatsoever, nor by way of trade or otherwise be lent, resold, hired out, or otherwise circulated in any form of binding or cover other than that in which they are published, without the publisher's prior written consent.

Notice: This book contains general information that is based on the author's own knowledge and experiences. It is published for general reference and is not intended to be a substitute for professional advice. The publisher and the author disclaim any personal liability, either directly or indirectly, for the information contained herein. Although the author and publisher have made every effort to ensure the accuracy and completeness of the information contained herein, we assume no responsibility for errors, inaccuracies, omissions and inconsistencies.

A CIP catalogue record for this title is available from the British Library.
ISBN 978-0-9928163-3-9

Cover design by Ian Lawton.
Cover photograph by Pawel Worytko, licensed by dreamstime.com.
Author photograph by Simon Howson-Green.

My sincere thanks go to Graham Williams of GSW Consultancy, one of the authoring team of the official PRINCE2 manuals, for reviewing the original manuscript of this book and making a number of important suggestions for improvement.

CONTENTS

INTRODUCTION	1
1 THE IMPORTANCE OF THE BUSINESS CASE	3
2 THE STRUCTURE OF PRINCE2	8
3 HOW THE ELEMENTS FIT TOGETHER	12
4 HOW THE PROCESSES FIT TOGETHER	19
5 FROM START TO FINISH IN OUTLINE	22
6 THE BUSINESS CASE THEME	28
7 THE ORGANISATION THEME	31
8 THE QUALITY THEME	36
9 THE PLANS THEME	40
10 THE RISK THEME	47
11 THE CHANGE THEME	52
12 THE PROGRESS THEME	58
13 STARTING UP A PROJECT PROCESS	64
14 INITIATING A PROJECT PROCESS	68
15 MANAGING PRODUCT DELIVERY PROCESS	72
16 CONTROLLING A STAGE PROCESS	74
17 MANAGING A STAGE BOUNDARY PROCESS	78
18 CLOSING A PROJECT PROCESS	81
19 BRINGING IT ALL TOGETHER	85
20 TAILORING PRINCE2	88
APPENDIX 1: PRINCE2 TAILORING QUESTIONNAIRE	95
APPENDIX 2: PRINCE2 PRINCIPLES CHECKLIST	98

FIGURES

Figure 1: The business case — 7
Figure 2: Overview of the four integrated elements — 18
Figure 3: PRINCE2 Line Diagram — 21
Figure 4: PRINCE2 Summary Diagram (in outline) — 27
Figure 5: The business case and benefits management approach — 30
Figure 6: The project management structure — 35
Figure 7: The quality audit trail — 39
Figure 8: The 3 levels of plan — 45
Figure 9: The product-based planning technique — 46
Figure 10: The risk management procedure — 51
Figure 11: The issue and change control procedure — 57
Figure 12: Setting/monitoring tolerances and escalating exceptions — 63
Figure 13: Starting up a project process — 67
Figure 14: Initiating a project process — 70
Figure 15: Managing product delivery process — 73
Figure 16: Controlling a stage process — 77
Figure 17: Managing a stage boundary process — 80
Figure 18: Closing a project process — 84
Figure 19: PRINCE2 Summary Diagram — 86

INTRODUCTION

PRINCE2 (*Projects in Controlled Environments*) is a leading, internationally recognised, project management methodology. It has been put together over several decades by committees of highly experienced project managers from many different fields.

PRINCE was originally developed for use in IT projects in the UK public sector only but, with the launch of PRINCE2 in 1996, the method was expanded to cater for any type of project and licensed for general use. For many years the role of accreditor was undertaken by APMG International, but in 2014 this was taken over by AXELOS, a joint venture between the Cabinet Office and Capita.

This guide is in no way intended as a replacement for the official PRINCE2 manual, but rather as a gentle introduction to it. The aim is to highlight the fundamental aspects of PRINCE2, and to provide simple explanations and diagrams. In particular those presented in the early chapters – which aim to show how the different elements of PRINCE2 fit together, and to provide a broad contextual understanding – are relatively unique in their presentation.

This guide should be particularly useful as pre-reading for anyone preparing to attend a PRINCE2 course, or as an introduction for those attempting self study, or for those who simply wish to gain a basic understanding of the method without necessarily becoming qualified in it. It does not focus on passing PRINCE2 exams. There are plenty of other publications that provide exam tuition, practice papers and so on. However the intricacies of taking and passing exams, combined with the slightly convoluted language sometimes used in the official manual, can tend to detract from the basic understanding of the method that this guide sets as its aim.

In terms of coverage, though, this guide does go through the majority of the method in reasonable depth. The only significant topics not covered are some of the tailoring guidance and detailed techniques discussed in the theme and process chapters of the official manual – particularly relating to the plans and risk themes – and the detailed contents of the various documents used to manage

INTRODUCTION

a project as per Appendix A. A proficient PRINCE2 project manager would need to acquaint themselves with these aspects.

Finally, at its heart good project management is, like most things, common sense. Many of us had to learn 'on the job'. However and wherever we use PRINCE2, we should never lose sight of the fact that we're mainly imposing some structure and perhaps standard terminology on what should be a core of common sense. So if at any point what we're doing stops being common sense, we're probably doing something wrong.

AUTHOR'S NOTE

This book is deliberately written in an informal style to try to make the topic less daunting and more approachable. The word 'we' is consistently used to reflect the fact that it's primarily written from the perspective of a project manager.

Single quotes are reserved for colloquial phrases or those of my own making, to distinguish them from PRINCE2 terms. Where it may not be obvious that the latter are being used they've been placed in italics, especially on first usage, or bolded. Underlining represents my own emphasis. Figures are usually presented at the end of the relevant chapter.

1 THE IMPORTANCE OF THE BUSINESS CASE

Amongst other things the business case tells us why we're undertaking the project and what its aims are. It is probably <u>the</u> most important document on a project. But before we can produce it, right at the outset we need to derive measurable estimates for the 6 key *objectives* of the project, most of which are then recorded in the business case.

THE SIX PROJECT OBJECTIVES

Quality <u>What</u> is the host organisation trying to achieve in terms of the final deliverable or output? Is it a new information system of some sort, or a new building, or a conference? Of course we'll be told something about this at the outset, but almost certainly we'll have to investigate further by talking to the *customers* or *users*. The problem is that often they only come up with relatively vague statements of what they want this final output to be and do, so these need to be turned into more measurable or specific *acceptance criteria* as early as possible.

Cost Given what the users want in terms of quality of final output, <u>how much</u> is that going to cost? Remember that as a project manager we don't have to be a 'subject matter expert' – and, if we're not, then as early as possible we should be looking to get some input from *suppliers* to provide at least a rough estimate of cost, rather than relying on pure guesswork or on arbitrary budget constraints. This is recorded in the business case.

Time Given what the users want in terms of quality of final output, <u>how long</u> is that going to take? Remember again that as a project manager we don't have to be a subject matter expert, and if we're not then as early as possible we should be looking to get some input from *suppliers* to provide at least a rough estimate of the timescale. This is also recorded in the business case.

Benefits In simple terms, what can the host organisation expect to <u>gain</u> from doing this project – or what benefits can they expect the final output to achieve for them? On many projects these will, or at least ought, to be expressed in financially measurable terms of

1 THE IMPORTANCE OF THE BUSINESS CASE

increased revenues, or reduced costs, or both. It is also often the case that these benefits will only be released <u>after</u> the project is completed – a new product range only starts to make more money once it's been developed and is on sale. So, for example, the organisation might expect a 10% per annum increase in revenues, or a 5% per annum reduction in costs, for the first three years after the project is completed. These estimates of benefits are again recorded in the business case.

Scope This dovetails with quality in making sure there's no confusion between users and suppliers about what will be included within the scope of the project and what will be regarded as outside its scope.

Risk All projects involve change, which introduces uncertainty, hence risk. This always needs to be managed. As early as possible we must identify at least any <u>major</u> risks to our project that could remove its business justification, which will again be recorded in the business case.

THE 'QUALITY-COST-TIME TRIANGLE': A 'NEGOTIATED EQUILIBRIUM'

All this assumes that we've been given a somewhat vague *project mandate*, which contains an idea of what we're trying to achieve but no real numbers for cost and time – and this is sometimes how a project begins.

At the other end of the scale, however, we may be given a mandate in which the cost and timescales have been fixed for us, and it may also be relatively specific in terms of the expectations of the final output. Often in this situation it doesn't take a full subject matter expert to work out that the costs and timescales appear to be on the low side, sometimes badly so. In diagrammatic terms, this would mean that the 'quality-cost-time triangle' in the upper part of Figure 1 was no longer 'in equilibrium' – the three sides would no longer be of the same length. Our reaction might therefore be, "I don't know how I'm expected to deliver all that for so little money and with so little time, but there's nothing I can do, so I'll just accept it."

This situation has been particularly prevalent in the economic climate of the late noughties and early tens, and some project managers tend to go into 'victim' mode at this point. Yet a skilled project manager should also be a good negotiator. Of course the political situation in our host organisation and our need to keep paying the mortgage and so on will dictate that sometimes we take on projects where the 'quality-cost-time triangle' is clearly not in equilibrium. But perhaps far more often than we might think there may be scope for negotiation, in particular between users and suppliers. If the sponsors insist that the cost and timescales are fixed, can we get them to agree to reduce the scope or quality of the final output? Or if the latter is fixed, can we negotiate more time and money? Can we at least get the triangle closer to, even if not in full, equilibrium?

The thing to remember is that no one gains from a project that is underfunded or too time-constrained. Something will almost certainly suffer, most probably the quality of the final output. Expectations will not be met, and the benefits may turn out to be far lower than had been anticipated. Note that diagrammatically if the cost and/or time sides of the triangle are shortened then the whole area within it – the benefits – becomes smaller.

The foregoing of course assumes that we have a pretty good idea of our final output. But on more 'evolving' projects our understanding of this may be relatively broad at the outset, and will only become more specific as we proceed. In such cases the equilibrium of the triangle only becomes paramount once we have specifics to work with.

THE ONGOING VIABILITY OF THE BUSINESS CASE

What do we mean by a 'viable' business case? It is one in which the benefits outweigh – or are greater than – the investment in terms of cost and time, and the level of risk being faced. This is represented in diagrammatic form in the lower part of Figure 1. That's why in simple terms we might think of the business case as a 'cost-benefit analysis'. Of course if the benefits are financially measurable this is easily determined.

We can make the assumption, then, that the business case is viable – or the benefits outweigh the costs – at the outset of the project,

1 THE IMPORTANCE OF THE BUSINESS CASE

otherwise we would never begin. But what then? Do we just leave the business case on one side? Has it now done its job? No. We must regularly come back to it and update our estimates to ensure that it remains viable. PRINCE2 projects are split into *stages*, and we do this at the end or *boundary* of each one.

We all know that, for a variety of reasons, estimates of time and cost tend to go up rather than down once a project is underway. In addition, especially on longer projects, we should revisit our estimates of benefits in case market or other conditions have changed. As a result of all this, do the benefits still outweigh our investment in terms of cost and time, and still justify the level of risk we're taking?

What if the benefits are not financially measurable? The viability of the business case is then more based on whether the non-financial benefits, and even the basic reasons for doing the project, are still in place; or whether anything has changed sufficiently – perhaps in terms of increased estimates of time or cost – to make the project no longer worth doing. But it's still important to revisit it and to ask these questions.

What if the business case does become nonviable for whatever reason? Unless it can be saved by a major change in scope, the host organisation should have the courage to shut the project down – that is, instigate a *premature close*. This isn't always done in practice because it takes courage to stand up and say that a significant amount of money has been spent without result – but the crucial point is that we're agreeing not to waste any further money on the project.

PRINCE2 MADE SIMPLE

The initial scoping of the project needs to be a 'negotiated equilibrium':

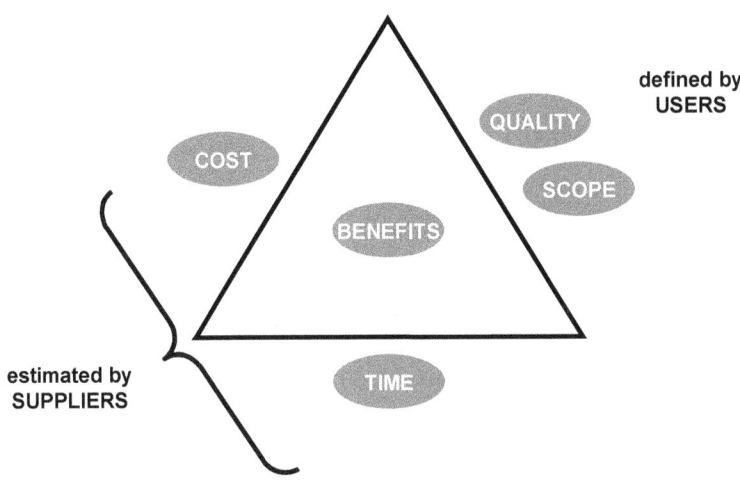

The business case needs to be updated at the end of each stage to check if it's still viable:

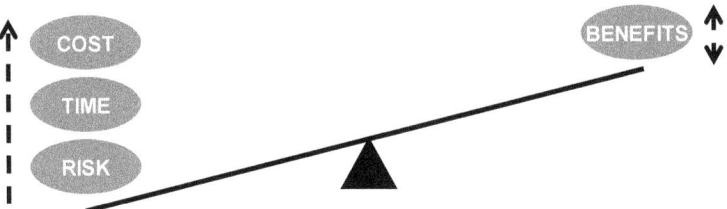

Figure 1: The business case

2 THE STRUCTURE OF PRINCE2

PRINCE2 is made up of four *integrated elements*: *principles, themes, processes* and *the project environment*. Let us look at each in turn.

1 THE SEVEN PRINCIPLES

These are the fundamental principles of good project management practice that should underlie everything we do:

- continued business justification
- defined roles and responsibilities
- manage by exception
- manage by stages
- focus on products
- learn from experience
- tailor to the project.

Crucially it's whether we're following all seven principles that most obviously determines whether or not we can say we're properly following the PRINCE2 method.

2 THE SEVEN THEMES

These are the various *disciplines* that should be adhered to throughout the project, i.e. when we're performing different activities, if we're to put the principles of best practice into effect (the abbreviations in brackets are used in various diagrams):

- business case (BC)
- organisation (ORG)
- quality (QUAL)
- plans (PLS)
- risk (RISK)
- change (CHGE)
- progress (PROG).

3 THE SEVEN PROCESSES

These are *sets of recommended activities* that to some extent provide a *timeline* for how we move through a project (again the abbreviations in brackets are used in various diagrams):

- ➢ starting up a project process (SU)
- ➢ initiating a project process (IP)
- ➢ managing product delivery process (MP)
- ➢ controlling a stage process (CS)
- ➢ managing a stage boundary process (SB)
- ➢ closing a project process (CP)
- ➢ directing a project process (DP).

4 THE PROJECT ENVIRONMENT

This comprises all the organisational influences on the project, such as established BAU processes, methods, standards and so on, as well as any industry-specific, legislative and other external factors that might be relevant. All this needs to be considered when we're deciding how to *tailor* PRINCE2 both to our organisation generally and to any specific project. More on this shortly.

MANAGEMENT AND SPECIALIST PRODUCTS

Although they're not one of the four integrated elements, PRINCE2 *management products* are immensely important to the method. These are the documents used to run any PRINCE2 project – in other words they're generic – and they're all defined in detail in terms of their purpose, composition and so on in Appendix A of the official manual. The business case, for example, is a management product, as are all the plans, progress reports, logs and registers, and a variety of other documents that we'll discuss in due course.

Another term that is immensely important in PRINCE2, which contrasts with these, is that of *specialist products*. These are the unique outputs or deliverables that we're producing or procuring on any particular project. They may be tangible, as in a new product range, or a new building; or they may be intangible, as in a new suite of software, a new set of processes or a reorganised department;

2 THE STRUCTURE OF PRINCE2

larger projects in particular may involve a combination of the two.

As an example, let's say we're building a new factory. The final output, which PRINCE2 refers to as the *project product*, is the fully fitted-out factory. The individual specialist products that go to make this up are – at a relatively high level of definition – the foundations, the shell, the internal and external walls, the roof, the production equipment, the offices, the fixtures and fittings, the toilets, the canteen, the car park and so on.

THE TAILORING OF PRINCE2

Have you ever heard someone suggest that PRINCE2 is an 'overly bureaucratic' method? That it involves always filling out every single document in as much detail as possible as often as possible? That it's all overkill?

PRINCE2 does have this reputation in some places, especially in certain parts of the UK public sector. However that's only because some people have never understood how to *tailor* it properly in different project environments. Tailoring has always been implicit in PRINCE2, but with the advent of PRINCE2 2009 it was made much more explicit. Not only was it given its own chapter in the official manual, but it also became one of the seven principles. Continuing this trend, tailoring considerations relating to each of the themes and processes were added in PRINCE2 2017, along with a new chapter on tailoring the method to the organisation as a whole. All this will be examined in chapter 20. The main environments considered for tailoring are:

- simple projects
- those that use an agile approach
- those that involve external suppliers
- those that involve multiple owners/sponsors
- those that are part of a programme.

All this means that for some time there's been no excuse for not understanding how to tailor PRINCE2. The most obvious factor that will influence the formality with which we use the method on any given project is its size or scale – measured in terms of the budget, timescale, number of people, complexity and risk, and so on. Although this will be relative to the size of the organisation hosting the project, in any one organisation we can be managing some projects that are regarded as small and simple, and some that are large and complex. It should be common sense that we would <u>not</u> want to adopt the same level of formality at each end of the spectrum.

We will see in chapter 20 that on a simple PRINCE2 project we can use just <u>two</u> people and <u>four</u> pieces of documentation. That proves it's emphatically <u>not</u> an overly bureaucratic method.

SO WHAT DOES 'FOLLOWING' PRINCE2 ENTAIL?

Although it's hugely flexible and tailorable, there are a minimum set of requirements for an organisation or a part thereof to be able to say it's running its projects according to the PRINCE2 method. These are:

- that it's applying <u>all</u> of the principles
- that it's meeting the minimum requirements laid down for each theme
- that it's ether using the recommended PRINCE2 techniques within each theme, or using alternative, equivalent techniques
- that its processes at least satisfy the purpose and objectives of those in PRINCE2, even if some activities are simplified, combined or even omitted.

3 HOW THE ELEMENTS FIT TOGETHER

Probably the most daunting aspect of PRINCE2 on first exposure is trying to work out how all these different *elements* fit together. This is what the 'Overview Diagram' in Figure 2 attempts to show at a very high level.

First of all, the *project environment* sits as an overarching element. As we saw in the previous chapter this comprises all the organisational and other external influences on the project.

Next, let us examine the most obvious relationships between the principles of best practice in the left-hand column, and the themes or disciplines by which they're put into effect in the middle one, as represented by the arrows between the two. Tailoring is of course a ubiquitous principle that can be applied to any theme or process, so it isn't considered separately below.

CONTINUED BUSINESS JUSTIFICATION

This is the principle, already discussed in chapter 1, that the business case must remain viable throughout, otherwise the project should be changed or shut down. It will be no surprise that this is fundamental to the **business case** theme.

However there's also a strong relationship between this principle and the **risk** theme. We've already seen that risk is all about uncertainty, which always exists on projects because they always introduce change of some sort. Major risks are things that could render our business case completely nonviable, and they should be recorded in it. Even on a smaller project there are likely to be one or two major risks of this sort. So if a business case hasn't identified any major risks, it's questionable whether it's viable at all.

DEFINED ROLES AND RESPONSIBILITIES

It should go without saying that on any project the people involved should have clear roles and be clear about their responsibilities within those roles. This principle lies at the heart of the **organisation** theme, in which the various roles and their associated responsibilities are defined.

It will be useful at this point to introduce the four levels in the project management hierarchy, each of which reports into and is accountable to the one above it:

- *Corporate, programme management or the customer* (CPC) act as the 'sponsors' of the project and provide it with its mandate. If it's part of a programme – that is, an umbrella for a number of related projects – then it's easy to understand that there will be a layer of programme management, usually a *programme board*, sitting at the top of the structure. But even when it's a standalone project there's often a layer of corporate management above the project board. Alternatively if we're looking at a project from the perspective of an external supplier, the host organisation is our customer.
- *The project board:* This is made up of three roles:
    - The *executive*, who has the controlling vote, represents the *business* or financial *interest* on the project and makes sure it delivers *value for money*. They are often regarded as part of corporate, programme management or the customer.
    - The *senior user(s)*, who represents the *user* interest. Within the project users are those people who define their requirements in terms of what they want the specialist products of the project to deliver for them, and then check these products when they're complete to ensure they're of the requisite quality. But in addition they may well use the products in some way to deliver the forecasted benefits, often only after the project is over.
    - The *senior supplier(s)*, who represents the supplier interest. Suppliers are those people or teams who create or procure the specialist products.
- *The project manager:* This role is responsible for the *day-to-day control* of the project and undertakes the vast majority of the activities within it, including preparing most of the

3 HOW THE ELEMENTS FIT TOGETHER

management products or documents.

➢ *Team manager(s):* Sometimes referred to as workstream leads, this role manages the *team members*. Typically these take the form of supplier teams developing specialist products, who can be thought of as 'at the coalface'. They might be programmers, or architects and builders, or any number of other specialists depending on the nature of the project. Supplier teams can be internal or external to the customer or host organisation, or a combination of the two.

MANAGE BY EXCEPTION

Many organisations use RAG (red-amber-green) reporting or 'traffic lighting' within progress reports. This is better than nothing but often it's up to the project manager, sometimes in conjunction with the project board, to use their own discretion as to what should be coloured amber or red.

Proper management by exception in PRINCE2 differs from this in two crucial ways. First, rather than discretion it uses defined, measurable *tolerances* to determine whether something is in *exception*. Second, these tolerances are set by the level of management above, establishing the level of authority that's being delegated. So, for example, the project board might say to us as project manager, "For the next stage we'll give you £200k +/- £10k, and 4 months +/- 1 week, and as long as you're forecasting that you can complete the stage within tolerance, send us monthly progress reports but there's no need for us to meet up." These are examples of cost and time tolerances, and they're arguably the most important ones at least on more traditional projects.

On the face of it, then, PRINCE2 has no amber warning, only red. But theoretically there would be no problem with building in an amber level of tolerance – e.g. up to +/- 2% is green, +/- 2% to 5% is amber, and over +/- 5% is red. In addition PRINCE2 progress reports always have a section for major issues and risks that may be of concern, even if we're not ready to go into exception yet.

As for the themes in which this principle is embedded, a key aspect of the **change** theme is how we handle project issues that come up

– such as a key specialist resource that's in short supply leaving or falling ill. The question then arises, what's the impact of this on our estimates of cost and time, and are we forecasting that it will cause us to exceed our tolerances?

Additionally, under the **progress** theme, we're regularly reviewing how we're getting on compared to our original plans, updating them with our 'actuals' in terms of time and cost, and reforecasting forwards to the end of the relevant plan. This exercise can, again, take us into exception.

MANAGE BY STAGES

We have already seen that a PRINCE2 project is split into *stages*. This principle is embedded within the **progress** theme because the *stage boundary* at the end of all except the final stage is a major control point at which our progress is assessed by the project board. In particular they're checking whether, after we've updated the overall project plan and business case, the principle of continued business justification is being maintained. These two management products are arguably the most important of all, even on a simple project, and can be thought of as two of the main 'drivers' of any project.

This principle is also built into the **plans** theme. Let us consider the fact that it's not possible to plan an entire project in detail at the outset if it's longer than, say, 3-6 months – although this figure is only the roughest of guides and will depend on the nature of the project and the environments in which the organisations involved operate. For that reason, if no other, it's sensible to split a longer project into stages so that we have realistic planning horizons for more detailed stage plans.

FOCUS ON PRODUCTS

Remember that in PRINCE2 the word *products* is interchangeable with deliverables or outputs, especially when we're considering specialist products. We saw in chapter 1 that the only way we can guarantee the correct outcome on our project, in terms of the quality expected, is by defining the measurable *acceptance criteria* we expect the final output or project product to satisfy. This in turn can only be achieved if each of the individual specialist products that go to make it up themselves satisfy their own measurable *quality criteria*. This is how this principle is embedded within the **quality** theme.

However it's less obvious as to why this principle is also embedded in the **plans** theme. This is probably the major area in which PRINCE2 differs from established practice, because it recommends the use of the *product-based planning technique* to identify and define the various specialist products before we start thinking about the activities that will produce them. Of course that's not to say that project managers don't think about the required deliverables at all when they're putting a project plan together, but they probably don't think about them as comprehensively and formally as PRINCE2 suggests they should, at least on a larger project. Nearly everyone who has ever used the *product-based planning technique* in practice, whether unwittingly or otherwise, reports that it was well worth the effort – and we'll find out why when we consider it in more detail in chapter 9.

LEARN FROM EXPERIENCE

Finally the fundamental principle of learning from the experience of previous projects, and also passing on our own learning experiences – both good and bad – to future projects, is built into the **quality** theme because it involves 'continuous improvement'.

As for the remainder of the Overview Diagram in Figure 2, the processes in the right-hand column provide us with a rough timeline through the project.

The starting up a project and initiating a project processes are used primarily by the project manager to get us off to a controlled start. Then in order to have a controlled middle the project manager spends most of their time using the controlling a stage process, in conjunction with the team manager(s) who use the managing product delivery process to get the real work done, i.e. specialist products produced or procured; the project manager then turns to the managing a stage boundary process just before the end of all except the final stage. To achieve a controlled <u>close</u> the project manager finally uses the closing a project process. From the end of start up onwards the project board use the directing a project process to exercise overall control.

The large bracket and arrows to the right of the middle column are intended to convey the idea that the themes are disciplines that should be adhered to <u>throughout</u> the project when the various processes are being used.

3 HOW THE ELEMENTS FIT TOGETHER

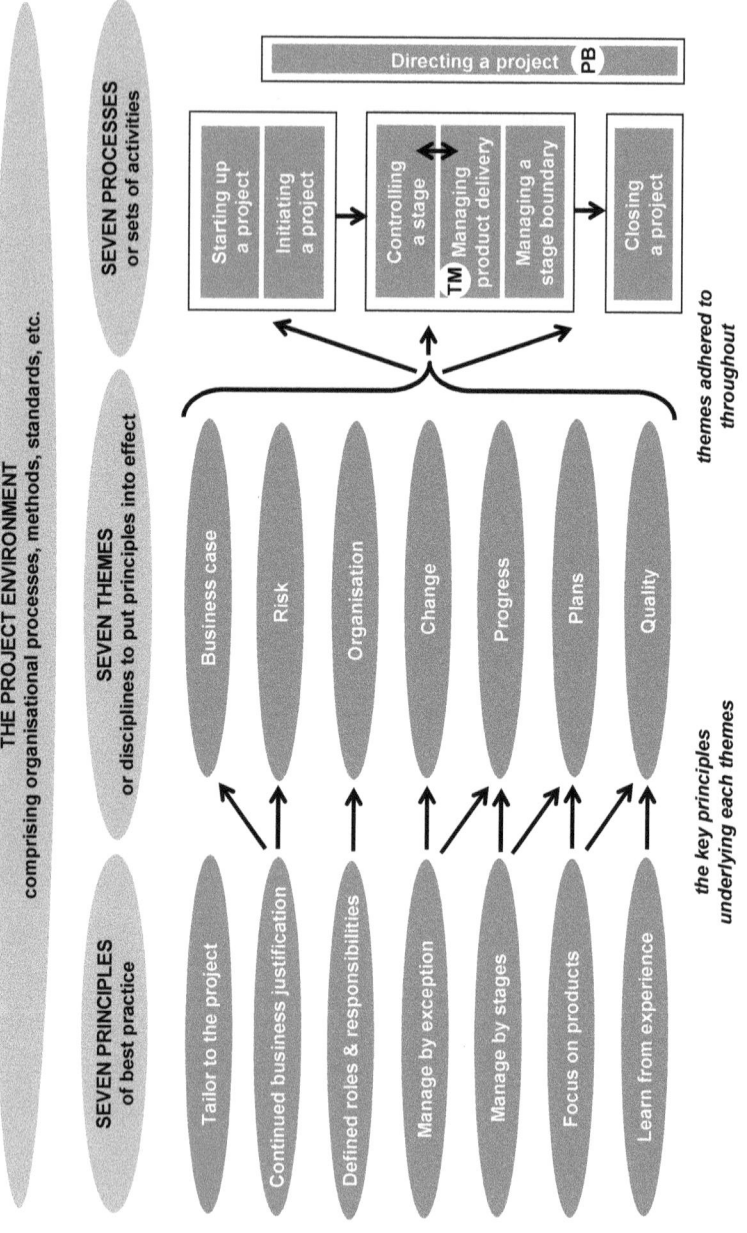

Figure 2: Overview of the four integrated elements

4 HOW THE PROCESSES FIT TOGETHER

The entirety of this chapter refers to the 'Line Diagram' depicted in Figure 3.

STARTING UP A PROJECT

Corporate, programme management or the customer trigger start up by issuing a **project mandate**. Why is it a separate process from initiation, and indeed regarded as pre-project work? Because it's something of a 'finger in the air' exercise in which the aim is to do the minimum necessary to see if it's worthwhile to undertake a full initiation of the project. In other words, we don't want to waste time and money performing a full initiation only to establish that the project isn't viable anyway. So we produce only an outline business case in start up, and this forms part of the **project brief**, the main management product output of the process. If start up is used properly and estimates of benefits aren't over-optimistic, many projects won't even get off the ground.

INITIATING A PROJECT

If the outline business case does look viable then the project board authorise us to proceed to initiation, which is always stage 1 of the project. In this process we're updating and refining the various parts of the project brief, and adding new elements to produce a full **project initiation documentation**, or **PID** for short, the main management product output of this process.

The remaining processes are shown in Figure 3 on the basis of an example project that has three further *specialist* or *delivery stages*: 2 Design, 3 Build and 4 Install & Train. These are not fixed stages in PRINCE2, which allows us to tailor the number of stages so there are as many or as few as required. They are simply being used as a typical example for illustrative purposes.

CONTROLLING A STAGE & MANAGING PRODUCT DELIVERY

The project manager spends most of their time on a PRINCE2 project controlling (the specialist) stages, working in conjunction with

4 HOW THE PROCESSES FIT TOGETHER

the team manager(s) who manage the delivery or procurement of the specialist products. These processes are used in combination nearly all the way through stages 2, 3 and 4.

MANAGING A STAGE BOUNDARY

Towards the end of stages 1, 2 and 3 the project manager will switch into managing a stage boundary. One of the key objectives here is to update the **project plan** and **business case** so that the project board can conduct an *end stage assessment*. This is held at the end of all except the final stage to assess whether the key principle of continued business justification is still in place.

Note that the splitting of stage 3 into parts A and B in Figure 3, and the insertion of an extra stage boundary between them, will be discussed in chapter 5.

CLOSING A PROJECT

At the end of the final stage the project manager will turn instead to the closing a project process. This is <u>not</u> a <u>stage</u>. It is simply a set of activities performed towards the end of the final stage, whatever that happens to be, instead of those of the stage boundary process. The reasoning behind this is that if there's a need for a premature close the <u>process</u> can be slotted in anywhere, unlike a fixed final stage. There is no end stage assessment here because we're no longer concerned with continued business justification.

DIRECTING A PROJECT

This process only officially starts when the first key project board decision is taken, i.e. to initiate the project at the <u>end</u> of start up. From then on it's used by them to provide overall direction throughout the project until its close.

PRINCE2 MADE SIMPLE

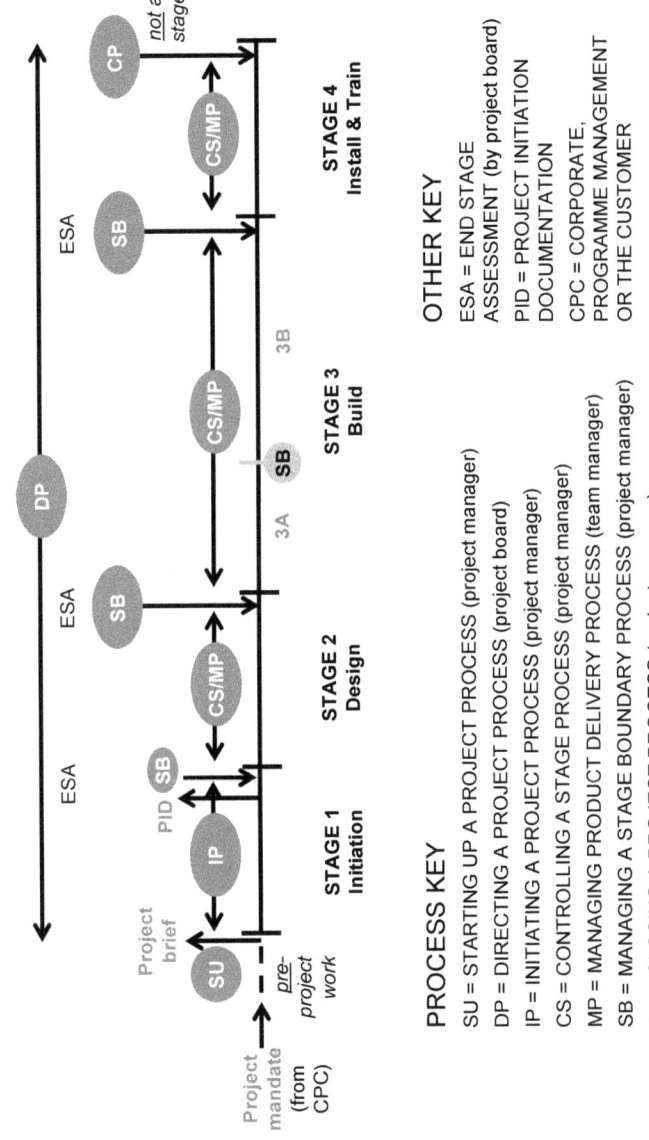

Figure 3: PRINCE2 Line Diagram

5 FROM START TO FINISH IN OUTLINE

Figure 4 contains a 'Summary Diagram' that builds on the Line Diagram discussed in the previous chapter. It again depicts how a project moves through the processes from start to finish, but this time with a somewhat different layout. It also brings in the most important management products involved in each process. Perhaps most unusually of all for any diagram attempting to summarise PRINCE2, it continues with the objective of showing how the different elements fit together by depicting which theme or discipline is being used when each management product is being created, updated or reviewed – except in that small number of cases where there is no obvious related theme.

To explain the layout of our unique Summary Diagram in more detail:

- The four layers of the management structure are represented on the left-hand side.
- The top row of boxes represents the individual activities in the directing a project process, the middle row represents the processes used by the project manager, and the bottom row the process used by the team manager(s).
- Management products are shown next to the various arrows and are bolded. Most of these are created or updated by the project manager, as represented by anything next to an arrow that exits either above or below the middle row of process boxes. Those few created or updated by the team manager(s) are next to the right-hand arrow exiting upwards from the managing product delivery box on the bottom row.
- Management products next to arrows that don't 'go anywhere' or connect to another box are being created or updated in that process but not being given to anyone else.
- (U) means a management product is being 'updated'.
- The themes are shown in grey ovals. The abbreviations used are as per the list in chapter 2.

> If a management product is shown with a list of bullet points underneath it, this means it's a composite document that includes these others within it. On a larger project it will be a compilation of separate documents.

> Any writing in grey relates to 'project not as usual' or *exception* situations.

As an introduction Figure 4 is just the outline version of our Summary Diagram, in that is depicts only some of the most important management products. It is intended to give us a broad contextual framework and understanding of a PRINCE2 project from start to finish, into which we can then insert further detail piece by piece. After we've examined all the themes and processes more thoroughly in the intervening chapters we'll pull everything together by completing our Summary Diagram in chapter 19.

STARTING UP A PROJECT

To begin at the beginning, corporate, programme management or the customer issue a **project mandate**, which is given to and appoints the executive on the project board. They in turn appoint the project manager who undertakes the rest of start up. The major output of this process is the **project brief**, a composite document that contains not only the **outline business case** but also a first draft of the **project product description,** which can be thought of as a high-level requirements or specification document for the project as a whole. It is also one of the main 'drivers' of any project because, as we saw in chapter 1, typically we must have some idea of the measurable *acceptance criteria* that will make our final output fit for purpose – otherwise we can have no idea as to whether our 'quality-cost-time triangle' is in any sort of equilibrium.

INITIATING A PROJECT

If the outline business case is viable the project board will *authorise initiation* – at which point both the project itself, and the directing a project process, officially start. The main output of initiation is the **project initiation documentation** or **PID**, another composite document. We transfer the various parts of the project brief into the PID and update or refine them as part of initiation – as represented

5 FROM START TO FINISH IN OUTLINE

by the arrow going between the two in Figure 4.

We also create various new management products in this process, by far the most important of which is the **project plan** – the production of which allows us to refine our estimates of cost and time in what will now be a full business case. These management products, remember, are the other two main 'drivers' of any project.

Then, provided the business case is still viable, the project board will take the decision to *authorise the project*. The naming of this activity can be a little confusing because the project officially started at the end of start up, but what is meant is that they authorise the *specialist work* to begin. At the same time we'll switch into a 'mini version' of the stage boundary process, mainly to produce a more detailed **stage plan** for stage 2 – which in the example in our Line Diagram in the previous chapter is the Design stage. This means the project board is taking the decision to *authorise a stage plan* simultaneously.

CONTROLLING A STAGE & MANAGING PRODUCT DELIVERY

Our first order of business when we start to control a specialist stage is to negotiate **work packages** with the relevant team managers. This will indicate how much time and money is allocated to each piece of specialist work. If there are multiple work packages in any given stage they may all be undertaken together from the start, or consecutively, or a combination of the two. Typically work packages will be undertaken by suppliers working within the host organisation, or by *external* suppliers in commercially separate organisations or, again, by a combination of the two.

While a team manager is executing a work package they will produce regular progress reports, typically weekly, called **checkpoint reports**. These allow us as the project manager to monitor their progress, but we in turn then produce regular **highlight reports** for the project board, typically monthly. While controlling a stage we can also ask the project board to *give ad hoc direction* about, for example, project issues or risks.

We will leave the discussion of the exceptional procedures shown in grey to the end of this chapter.

MANAGING A STAGE BOUNDARY

When we're getting near the end of stage 2, if there are further stages as in our example we'll switch to managing a stage boundary. Our main purpose here is to update the **project plan** with our progress to date and with revised forecasts for the remainder of the project, to use these to update the overall time and cost estimates in the **business case**, along with any revisions to our forecasted benefits, and to build these updates into an **end stage report** on our progress to this point. This and a **stage plan** for stage 3, in our example Build, are presented to the project board so they can hold an end stage assessment to see if we have continued business justification, in which case they'll *authorise the stage plan*.

At this point we'll switch back and start to control stage 3, negotiating work packages, receiving checkpoint reports, producing highlight reports, then at the end of stage 3 we'll switch to managing a stage boundary again. If the updated business case is still viable then the plan for stage 4, in our example Install & Train, will be authorised, and we'll switch back to controlling this final stage.

CLOSING A PROJECT

When we get towards the end of stage 4 we'll switch into closing a project process and produce our final progress report for the project board, the **end project report**, so they can *authorise project closure*. This will represent the official close of the project, which needs to be a clearly defined point at which everyone involved understands that we've switched from a project/development to a BAU/live/ongoing maintenance environment, and the project management team has been disbanded.

EXCEPTION PROCEDURES

If we now refer to the grey elements in Figure 4, let's say that around a third of the way through stage 3 we revise our cost forecast to a total of £215k to complete the stage, which would take us beyond our cost tolerance of £200k +/- £10k. At this point we must escalate this to the project board via an **exception report**. This might happen because of a major project issue or risk, i.e. within the change or risk themes, or simply because we're re-evaluating our progress.

A relatively minor exception like this is very unlikely to threaten the overall project cost tolerance or the business case, so the project board will almost certainly want the project to carry on. They will ask us to produce an **exception plan** to replace the stage plan, and to do this we force an 'exceptional stage boundary', i.e. one that we didn't originally expect to have. If we refer back to Figure 3 we'll effectively split stage 3 into part A (the third already completed) and part B (the remainder) and we'll be primarily replanning for stage 3B.

Let us now change our scenario to one in which we have an overall project budget of £500k +/- £25k, and we're forecasting a cost overrun of £30k, but there's £80k of profit in the business case. In this case the business case is still viable so again we'll carry on and produce an exception plan, but this time it will replace the entire project plan and we'll need to get more budget from corporate, programme management or the customer.

Let us finally change our scenario to one in which we're forecasting a major cost overrun of say £100k, which renders the business case completely nonviable. In this instance the project board, in conjunction with corporate, programme management or the customer, will most likely recommend a *premature close* – unless a major change in scope, usually a reduction, can rescue the business case.

It would also be possible to have a premature close at a normal stage boundary when the project board might refuse to *authorise a stage plan* because of a lack of continued business justification – especially if, for example, the boundary was deliberately chosen to coincide with a major risk point, and that risk materialised despite our best efforts. Nor would it be impossible for this to happen in the mini stage boundary at the end of initiation, if our refining of the business case had shown it to be no longer viable.

PRINCE2 MADE SIMPLE

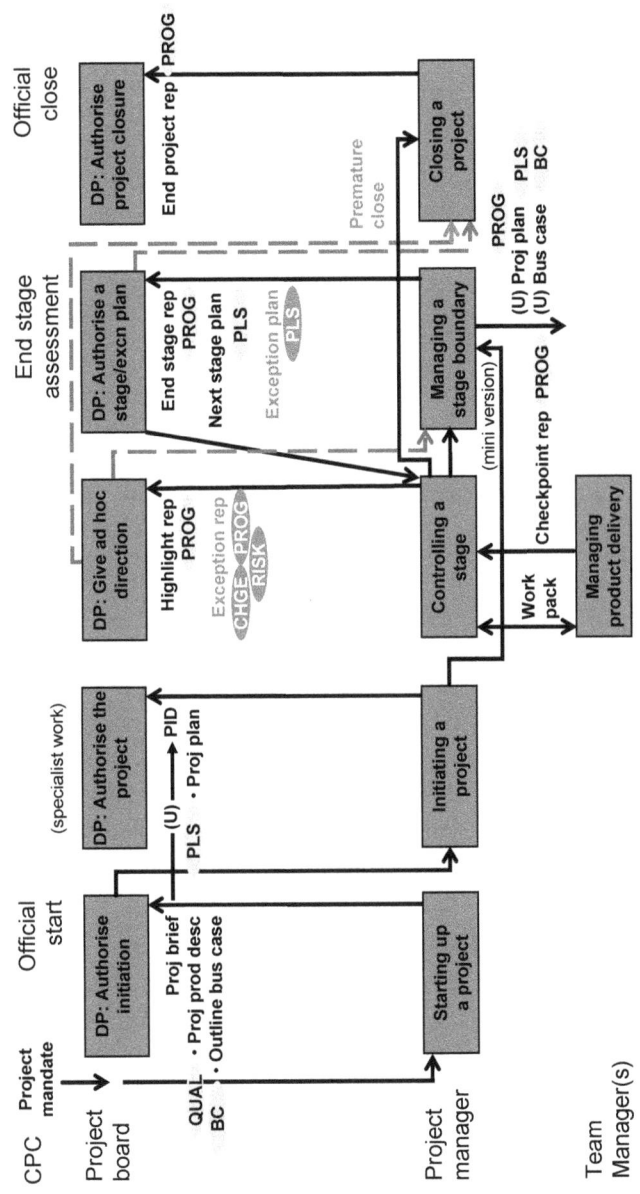

Figure 4: PRINCE2 Summary Diagram (in outline)

6 THE BUSINESS CASE THEME

[Note that in the detailed diagrams relating to the individual themes in this and the following chapters, management products or other important documents or parts thereof that are usually written down in some fashion are shown in grey and bolded, while processes are shown in dark grey circles and abbreviated as per the list in chapter 2.]

THE BUSINESS CASE

If we refer to the bottom part of Figure 5, in the previous chapters we've discussed at some length the development of the outline business case in start up (which, remember, is pre-project work) and of the detailed business case in initiation, and its ongoing updating or maintenance at each stage boundary. So this element of the diagram should be easy to understand.

Note also that there can be different types of business case depending on the motivation for doing the project. Much of the time we tend to assume that projects are financially motivated by increased revenues, reduced costs or both. But we can be involved in not-for-profit projects, such as infrastructural ones, or in compliance/legislative projects where the major aim is, often, not to incur a fine or a damaged reputation. Whatever the motivation we should always have a business case of some sort.

The typical contents of a business case would include: the reasons for doing the project; the options considered in terms of the project product; the expected benefits and disbenefits; the timescales and costs and any tolerances around those; an *investment appraisal* or 'cost-benefit analysis'; and the major risks.

THE BENEFITS MANAGEMENT APPROACH

Turning to the top part of Figure 5, we can see there the phrase *confirm benefits*. This involves actually checking to see whether the benefits originally forecast in the business case, which formed the basis for project approval, are delivered (or realised or actualised) by the project product.

To see how this works, let's return to the example used in chapter 1 in which we suggested that the host organisation is expecting a 10% per annum increase in revenues for the first 3 years after the project is completed – for example because they're launching a new product line. In this instance the increased revenues will only come on stream <u>after</u> the project is over, and once the new product goes on sale. So we might conduct a series of *benefits reviews* post project, for example at the end of year 1, year 2 and year 3, to see if revenue has actually increased by 10% in each. If the reality is more than 10% the host organisation will be even happier. But if it's less, especially significantly less, we ought to find out why. It could be because market conditions changed in a completely unexpected way, which is nobody's fault. But it could mean that whoever came up with the forecast estimates of benefits in the original business case significantly overestimated them by being way too optimistic – just the same way that any of us in our private lives tend to overestimate how much we might get when we sell our house or car, for example.

The senior user(s) is the PRINCE2 role assumed to be responsible for the estimates of benefits, on the basis that we'll need their buy-in that they can realise the benefits, which is what's crucial to overall success. Here it might be the marketing director, for example. So one function of the benefits reviews is to learn from experience and hold the senior user(s) to account, the aim being to make sure that in future estimates of benefits in business cases are more realistic. The problem if they're not is that many projects may be given the go ahead based on unrealistic business cases that should actually have been thrown out.

The key management product relating to this is the benefits management approach. It should be created during initiation for two reasons. First, because it's important to focus at the outset on how benefits will be measured, who will undertake the work and how much it will cost. Second, because we might have 'in-project' benefits being realised before it's over. For example, if we were installing a new system in a number of different locations and treating each one as a separate stage, the earliest implementations would start to release benefits before the end of the project.

6 THE BUSINESS CASE THEME

The benefits management approach is then handed to corporate, programme management or the customer at the end of the project because at this point the project management team will be disbanded, so they're the only people left to make sure the post-project benefits reviews get carried out.

THE IMPORTANCE OF OUTCOMES

In many cases the project must not only deliver the appropriate *outputs* or products, but also the appropriate *outcomes* – or 'organisational changes' – needed to make sure the benefits are realised. For example, let's say we're implementing a new document management system. The benefit we hope to achieve is to reduce costs by say 2%. But we'll only do this if the appropriate outcomes are achieved first – that is, documents are accessed more swiftly, and mistakes arising from using incorrect versions are reduced. It is vital that during the project itself we focus not just on creating the outputs but also, in this example, on the ease of use of the system, on appropriate staff training and post-project support, and on gaining proper commitment to use the system from all key stakeholders.

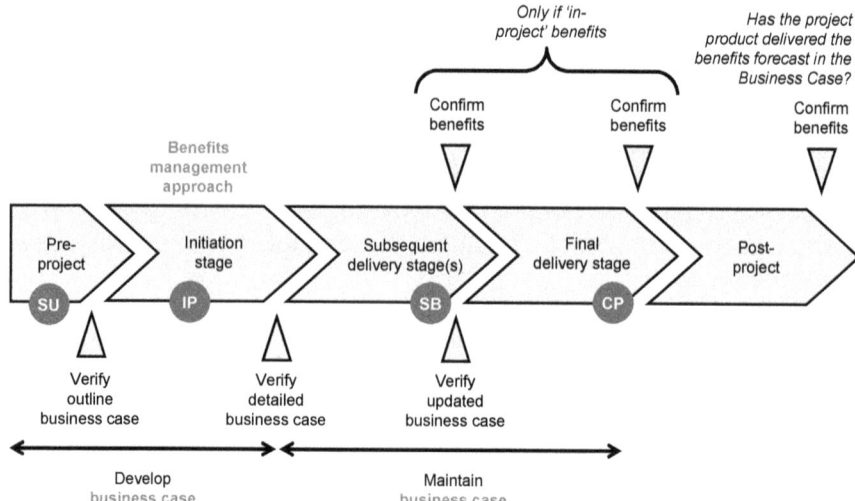

Figure 5: The business case and benefits management approach

(based on Figure 6.2 Managing Successful Projects with PRINCE2® 2017 Edition)

7 THE ORGANISATION THEME

THE ROLES IN THE PROJECT MANAGEMENT STRUCTURE

We have already discussed the core roles in the project management structure at the beginning of chapter 3. To recap and add a few more details, and referring now to Figure 6 as well, **corporate, programme management or the customer** act as the sponsors of the project and provide the mandate. Under them sit the **project board**, which may or may not be made up of separate people, and on which the three *project interests* are represented:

> - The **executive**, who is the ultimate decision maker, represents the *business* or financial *interest* and makes sure it delivers value for money.
> - The **senior user(s)**, who represents the *user interest*, i.e. those people who define their requirements in terms of the specialist products and then check them when they're complete to ensure they're of the requisite quality. There may be more than one person in this role, and they must be senior enough to commit the user resources the project requires.
> - The **senior supplier(s)**, who represents the supplier interest, i.e. those people who will create or procure the specialist products. There may be more than one person in this role, and they must be senior enough to commit the supplier resources the project requires.

Into them reports the **project manager**, who undertakes the vast majority of the activities, prepares most of the management products and is responsible for the day-to-day control of the project. They will normally come from the host organisation, although they can come from an external supplier. Into them reports the **team manager(s)**, who in turn manage the **team members** who do the specialist work. Because they might be programmers, or architects and builders, or any number of other specialists depending on the nature of the project, we can't be proscriptive about the work they do. Therefore, although their work 'at the coalface' is vital to the project, they're effectively operating underline of PRINCE2. Supplier teams can be

7 THE ORGANISATION THEME

internal or external to the host organisation, or a combination of the two.

In addition to these core roles we have **project support**. Many organisations have a centralised project management or project support office, who not only provide administrative support but can also be a source of expertise in, for example, change and version control (in PRINCE2 the latter is referred to as *configuration management*), risk management and planning.

Then we come to **project assurance**. Operating on behalf of the project board it typically asks the question: "Are things really going as well as we're being told?" For example, a project manager might be scared to admit in their highlight report that they're starting to encounter delays and cost overruns, hoping that these will be reversed in the next period – when in fact the situation will almost certainly get worse. The key phrase is that project assurance undertakes *independent monitoring*, not only of the project manager but also team manager(s) and project support, i.e. of everyone below the 'line of independence' superimposed on Figure 6. This role can actually be split into three separate components, that is *business*, *user* and *supplier assurance* conducted on behalf of each project board member respectively. Board members can do their own assurance, or they can delegate it in part or in full to separate people. Project assurance can also, in its 'good cop rather than bad cop' guise, provide advice and guidance to the project manager.

The final role is that of the **change authority**, which will be discussed under that theme in chapter 11.

ROLES NOT PEOPLE

The roles contained in the light grey box in Figure 6 are within that subset of the structure referred to as the *project management team*, and as it says at the bottom they're 'roles not people'. This means the team can be tailored to different sizes of project by one role being *shared* by more than one person, or by multiple roles being *combined* so that just one person fulfils them – even if the responsibilities attached to each role must always be allocated to someone.

The other fact to be aware of is that two of the roles are so crucial in terms of accountability that there must always be 'one and only one' person in each of them – these being the executive and project manager. In fact on a simple PRINCE2 project these two represent the minimum number of people who can carry out all of the roles: the executive taking on the responsibilities of the other project board roles and doing all their own assurance, i.e. everything above the 'line of independence'; and the project manager taking on the responsibilities of the team manager by working directly with just one or two team members, and doing all their own support, i.e. everything beneath the line.

As for appointing separate people in certain roles, firstly project board members will tend to do this with their aspect of project assurance if they don't have the requisite skills or enough time. Second we as the project manager might appoint separate people as team manager(s) for a variety of reasons: if it's generally a larger project with many teams and members; if we aren't a subject matter expert in all the different aspects of the project; and if we have multiple teams in widely spread geographical regions, so that it's easier to get a representative from each location to come and meet with us e.g. once a week rather than us having to visit them.

THE SIZE OF THE BOARD

The project board are there to act as a 'decision-making body not a talking shop'. For example as a project manager at each boundary we need them to authorise us to carry on with the next stage, with minimal delay. If a project board is overcrowded with people who have little of use to contribute to the project – and indeed might be actively and unnecessarily intrusive and meddling, in their desire to justify their sitting on it – it won't operate effectively. For this reason PRINCE2 suggests its size should be kept as small as possible.

MEETINGS WITH THE BOARD

It would be fine, although not essential, for the project board to meet with the project manager at a stage boundary. But <u>within</u> a stage, if the project manager was having regular e.g. monthly meetings with the project board <u>only</u> to discuss progress as per their highlight reports, that would be in direct contravention of the fundamental principle of *management by exception*.

However in practice there are a number of reasons why project boards or project managers can insist on regular meetings. Most important and arguably PRINCE2 compliant of these is if genuine decisions are being made in them, most obviously about major issues and risks, because arguably then the project board's *ad hoc direction* is being condensed into a monthly meeting that can be cancelled if there's nothing to discuss. It is also arguable that if project board members are doing their own assurance then they might want to have regular meetings. Less PRINCE2 compliant, although perhaps understandable, is when project managers find that their project boards simply ignore highlight reports if they're only emailed. Finally, of course, sometimes project board members simply don't trust the project management process and insist on micromanagement.

STAKEHOLDERS

A stakeholder is defined as *someone who can affect or (perceive themselves to) be affected by a project*. This of course includes not just corporate, programme management or the customer, but all the members of the project management structure. Team members too are often overlooked but can surely have a major impact if they're not motivated. Nor should we ignore people who are not just outside of the project management structure but outside of the host organisation completely – members of the public, for example, in local council initiatives; or people in government departments or regulatory bodies on some compliance-type projects.

Having identified and analysed our stakeholders in terms of their level of interest and influence/impact, we decide how we'll communicate effectively with them – most obviously by sending them progress reports. But this should be a two-way process where we sometimes obtain information from them as well. All of this analysis is documented in the **communication management approach**.

PRINCE2 MADE SIMPLE

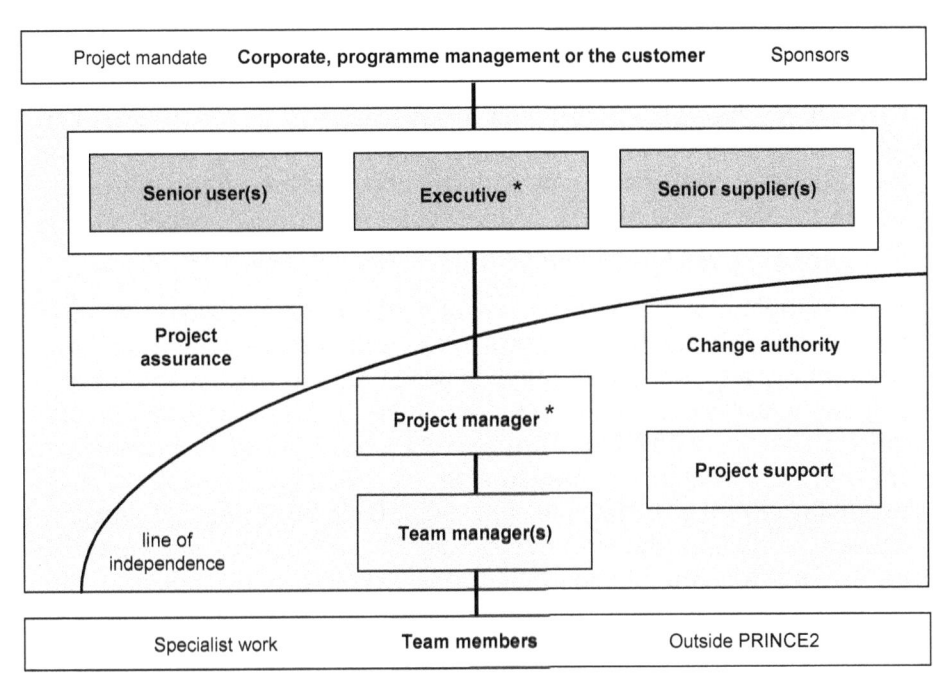

Project mandate	Corporate, programme management or the customer	Sponsors

- Senior user(s)
- Executive *
- Senior supplier(s)
- Project assurance
- Change authority
- Project manager *
- Project support
- Team manager(s)
- line of independence

Specialist work	**Team members**	Outside PRINCE2

* one and only one person in each role

these are *roles* not *people*

Figure 6: The project management structure

(based on Figure 7.3 Managing Successful Projects with PRINCE2® 2017 Edition)

8 THE QUALITY THEME

THE QUALITY-RELATED MANAGEMENT PRODUCTS

We saw in chapter 1 that it isn't always easy to tie down what the users within the customer or host organisation want in terms of the final output or deliverable. As early as possible the project manager needs to get an initial idea of the *customer's quality expectations*, but these may tend to be expressed in relatively vague terms.

For example, let's say that our project involves building a new factory, and initially the key users tell us that they need it to be 'big', 'state of the art' and 'professional looking'. Can we establish an even vaguely accurate estimate of cost and time to build the factory based on such information? No. They need to be pressed to provide us with more specific or measurable *acceptance criteria*. In this example 'big' might translate into '80,000 units a week production capacity', 'state of the art' into 'production line equipment model X from German manufacturer Y', and 'professional looking' into 'smoked-glass windows at the front, a water fountain near the entrance, and a tarmac-laid car park with 100 bays to one side'... and so on.

Headings for both customer's quality expectations and acceptance criteria are built into the **project product description**, as shown at the top left of Figure 7. As we saw in chapter 5 we can think of this as a high-level requirements or specification document for the project as a whole, and one of the main 'drivers' of any project. A first draft needs to be produced in start up to allow us to put rough estimates of time and cost into the outline business case, and then it will be firmed up during initiation.

Also during initiation we'll use the *product-based planning technique* for the first time to identify at least the major individual specialist products that will go to make up the project product. We will discuss this technique in more detail in chapter 9, but for each of these we'll then create an individual **product description** – which can be thought of as more detailed requirements or specification documents. As shown on the right of Figure 7, among other things these might contain information on the measurable *quality criteria* the product needs to satisfy; any *quality tolerances* around these, e.g. it must weigh 300g +/- 10g; the *quality method* to be used to check it when it's complete, e.g. review, physical inspection, walkthrough etc.; and the *quality responsibilities*, i.e. the *producer* of the product, the *reviewers* who will check it, and the *approvers* who will sign it off if they're different people.

Another management product relating to the quality theme that's created during initiation is the **quality management approach**. There are 5 of these documents – we discussed the benefits management approach in chapter 6, and mentioned the communication management approach briefly in chapter 7, but there's also one for risk and one for change control – and they define the practices and procedures we want to adopt in each of the 5 areas. For example, if a project involved many different specialist teams from many different organisations working on many different work packages involving many different specialist products, coordinating everyone's approach in the areas of benefits measurement, quality planning/control, risk management, change/version control and stakeholder engagement would be immensely important. However managing these aspects is far simpler on a smaller project involving few products and people. The quality management approach in particular describes the procedures, techniques and standards that we'll use to effectively achieve the required quality levels on the project, and the various responsibilities around them.

Then we come to the **quality register**. On larger projects involving many specialist products it's extremely useful for us to have a simple way to make sure they're all approved at the end of each stage. So in simple terms this is a 'diary of quality checks' that lists those checks or tests that will be performed on each product within each stage. This is the management product that overlaps between the *quality planning* undertaken by the project manager (see the large bracket on the top left of Figure 7) and the *quality control* that is the responsibility of the team manager(s) (see the smaller bracket on the bottom right – we'll discuss this latter aspect more in chapter 15). The planned checks for the next stage are inserted by the project manager while they're planning it as part of the stage boundary process, then the results are input by the team manager(s) as part of managing product delivery – although if separate people are undertaking project support they will usually be in charge of the actual updating of the register.

Finally, if we refer to the bottom left of Figure 7, when we're closing a project we should make sure we obtain written **acceptance records** from the operational team who will use/maintain the project product.

QUALITY ASSURANCE

Quality assurance is the general function within an organisation that establishes and maintains its quality management systems, concerning e.g. ISO and BSI compliance and so on. Sometimes this role is approximated by an 'internal audit' or 'compliance' function.

Quality assurance is <u>external</u> to any project, while project assurance is a role <u>internal</u> to the project management team. They both involve *independent monitoring*. But while quality assurance typically monitors compliance with organisational standards on behalf of corporate, programme management or the customer, project assurance typically monitors compliance with e.g. PRINCE2 standards of highlight reporting on behalf of the project board.

PRINCE2 MADE SIMPLE

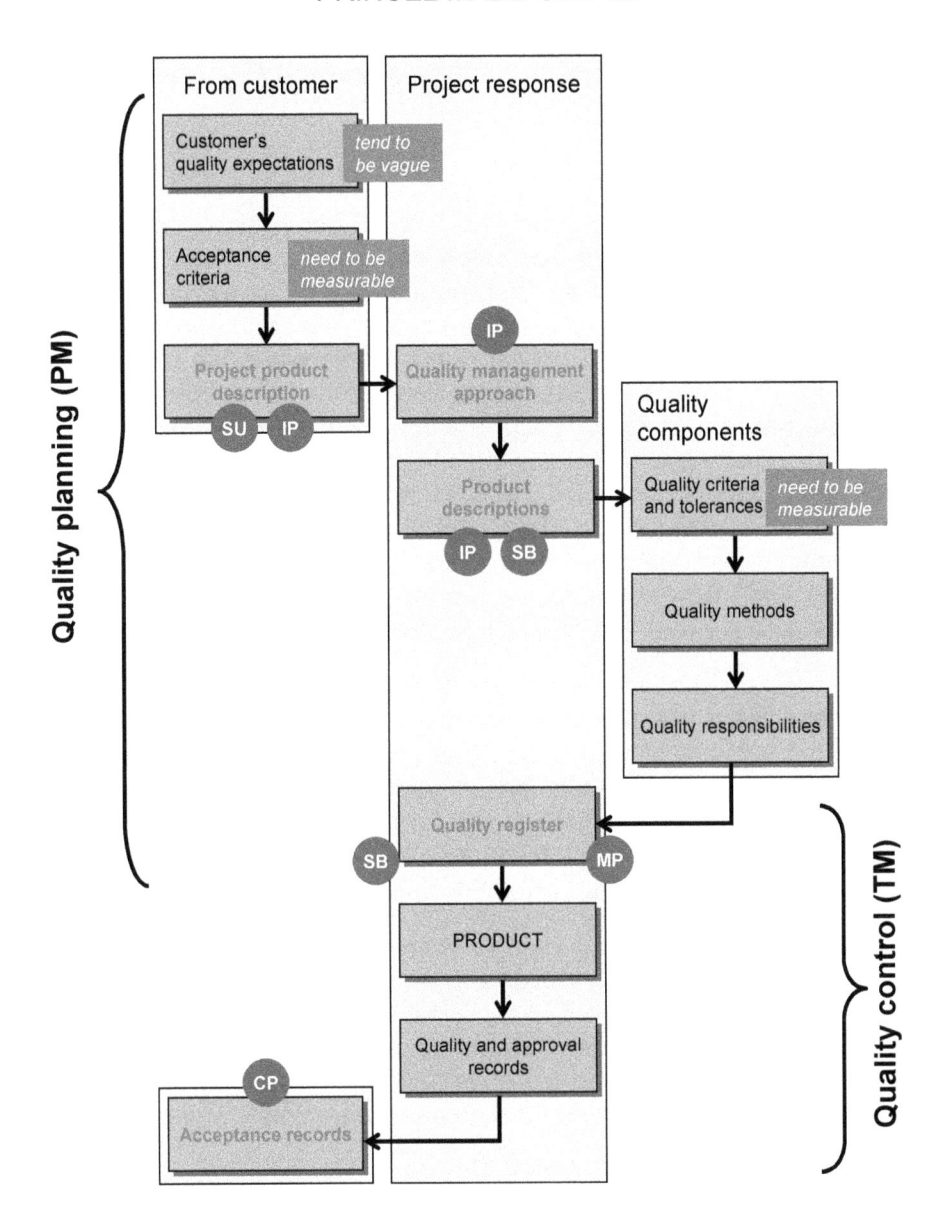

Figure 7: The quality audit trail

(based on Figure 8.1 Managing Successful Projects with PRINCE2® 2017 Edition)

9 THE PLANS THEME

THE THREE LEVELS OF PLAN

In chapter 3 we discussed the fact that it's not possible to plan an entire project in <u>detail</u> at the outset if it's longer than, say, 6 months – although this figure is only the roughest of guides and will depend on the nature of the project and the environment the host organisation operates in. For that reason, if no other, it's sensible to split a longer project into stages so that we have realistic planning horizons for more detailed stage plans, which are primarily used by the project manager for *day-to-day* control. However there is a third, even-more-detailed level of plan, a team plan. These are only optional but if required would be created by the team manager(s) in the managing product delivery process, typically to show how a work package will be executed – therefore with an even shorter timeframe.

The various processes in which the three different levels of **project, stage and team plans** are created are shown in Figure 8. As for the box at the top, if a project is part of a programme it's not difficult to understand that there would be an overall, high-level, programme plan of which our project plan would form a subset. If it's not part of a programme there won't necessarily be some sort of higher-level corporate or customer plan, but there might be.

Note that plans aren't only used to work out what needs to be done ahead of time, but they also act as a *baseline* against which progress can be monitored. So project plans will be updated with actual performance at each stage boundary, while more detailed stage plans will be updated while we're controlling a stage – typically daily or weekly, depending on the size and complexity of the project.

EXCEPTION PLANS

At the end of chapter 5 we saw that, if we're forecasting that we can't complete a <u>stage</u> plan within e.g. cost or time tolerance, the project board will ask us to produce an exception plan to replace it, to show how we intend to recover the situation.

But we also saw we could be forecasting that we're going to exceed the tolerances in the overall project plan, although not by so much that the business case becomes nonviable. In that case the project board would need to discuss the situation with corporate, programme management or the customer, and will almost certainly ask us to produce an exception plan to replace the overall project plan.

By contrast if there's only a threat to the tolerances at the work package or team plan level, this is resolved as a project issue and doesn't require an exception plan – as shown by the absence of dotted lines on the bottom right-hand side of Figure 8.

THE NUMBER AND LENGTH OF MANAGEMENT STAGES

Stage boundaries are crucial decision points on a project, at which the project board is assessing whether or not we have *continued business justification*. They act as a form of 'firebreak' to ensure that the project doesn't just steam ahead like a runaway train, without anybody ever asking whether we actually want to carry on.

There must be a minimum of two stages even on a simple project: initiation and the main delivery stage. But on larger projects our decision about what stages to have may be driven by the nature of the delivery work – in Figure 3 we used a simple example in which there were 3 further stages: Design, Build and Install & Train. However there are other factors that can influence our decision:

➤ If any stage would be longer than say 3-6 months we might split it up just to have a sensible time horizon for detailed planning.

➤ If there was a major risk point on a project it would be sensible to make that a stage boundary so we have an easy premature-close point. For example, if we were going ahead with a risky project now to steal a march on potential competitors, but some legislation was being discussed halfway through the Design stage that, if not passed, would render our business case nonviable, we'd split that stage in two. In other words 'business factors should override technical ones when deciding what stages to have'.

9 THE PLANS THEME

> ➤ In those periods of a project that are particularly complex and risky we might choose to have more and shorter stages, therefore more boundaries to keep tighter control – and we'd probably tighten up our tolerances on these stages too.

The other problem is that, with the increasing popularity of iterative approaches such as agile, we often find that delivery steps – such as design and build – overlap. In such cases it's still vital that our <u>management</u> stages do <u>not</u> overlap. That is because we must have definitive management stage boundaries at which all products have been approved, so that they can act as clear points at which our progress to date can be used to forecast forward for the rest of the project – whereas if one or more major work packages overlapped a stage boundary we'd have no clear idea of our progress at that point. For this reason it's vital that the delivery steps are somehow divided up so that we still have clear management stage boundaries. We will return to this topic in chapter 20.

CONVENTIONAL ACTIVITY-BASED PLANNING

Project managers unfamiliar with PRINCE2 will tend, when planning a project, to start by preparing a Gantt chart – a diagram/schedule containing a number of lines with defined start and end points, each one representing an activity or task, set against a horizontal timescale.

On simple projects MS Excel can be used to produce a simple schedule, but using more advanced software such as MS Project allows us to input estimates of how long each activity will take, the number of units of each resource type needed for each activity, their unit cost and extent of availability to the project, and so on. It will also allow us to determine the *critical path*, which represents those activities that, if delayed, push the remaining activities on the critical path out so that the whole project is delayed – whereas all other activities that aren't on the critical path can at least to some extent be shifted backwards or forwards without affecting other activities.

The activities on the critical path are potential risks to the project, and we might sensibly try to de-risk the plan by building some time slack into them, even if this isn't explicitly disclosed. The other major risks we will tend to identify when planning are critical specialist resources that are in short supply, any major assumptions we've made – for example about the cost and availability of resources – and dependencies on products from other parallel projects.

PRINCE2 regards a documented plan as being far more than just a diagrammatic schedule. Anyone who's been asked to review a Gantt chart of any complexity, which they haven't had a hand in producing, will know that one's eyes start to glaze over very quickly. So we'd expect to see not only the various schedules and diagrams in a plan, and estimates of time and cost and any tolerances around them, but also plenty of narrative – describing its scope in terms of the major products it covers and who's going to produce them, any assumptions we've made in producing it, any risks, any dependencies, and so on and so forth.

PRINCE2 PRODUCT-BASED PLANNING

We saw in chapter 3 that the PRINCE2 approach to planning is probably where it differs most from typical established practice because, based on the *focus on products* principle, it recommends the use of the *product-based planning technique* to identify and define the various specialist products we'll need before we start thinking about the activities that will produce them. Remember that's not to say that project managers don't think about the required deliverables at all when they're putting a project plan together, but they probably don't think about them as comprehensively and formally as PRINCE2 suggests they should on a larger project. Nearly everyone who's ever used this technique in practice, or something like it, reports that it was well worth the effort. In the next section we'll find out why.

If we refer to Figure 9 we can see there are 4 steps in this technique. The first is to write the **project product description** to define the project-level quality, while the third is to write the individual product descriptions to define the product-level quality. We have already discussed both of these, and the processes in which they're created or refined, at some length under the quality theme in the previous chapter, but they're also relevant to this technique.

9 THE PLANS THEME

The second step involves creating a **product breakdown structure**. This is a diagram in which we place the project product or final output of the project at the top and attempt to break it down into its individual components (an example is provided in Appendix D of the official manual, but we need not concern ourselves with it here). In practice this is best produced by putting key stakeholders representing both the user and supplier interests in a room with a white board and some sticky post-it notes and thrashing out what's going to be in scope and what out of scope. Such a meeting may require careful chairing to prevent it getting out of hand, but it can act as a hugely constructive negotiation between the two sides that not only builds consensus but also increases buy-in from key stakeholders. As such it can play a major role in making sure that the 'quality-cost-time triangle' discussed in chapter 1 is in equilibrium at the outset.

Returning to step 3, as soon as each product is <u>identified</u> as being needed in the product breakdown structure, then we should write at least a first draft of its **product description** as soon as possible afterwards. The final step in the technique is to create a **product flow diagram**, in which the products are put into the order or *sequence* in which they'll be procured or produced (again an example is provided in Appendix D of the official manual, but we need not concern ourselves with it here).

In terms of <u>when</u> all this is happening, a product breakdown structure would first be drawn up during initiation when preparing the project plan. However if this was a lengthy project then we wouldn't be able to identify <u>all</u> the individual specialist products required at the outset, only the major ones. In this case we'd need to revisit it when planning each stage in more detail in the stage boundary process, and possibly take it down to a lower level or even create a whole new subset structure just for that stage, creating some new product descriptions at the same time. On a larger project with separate team plans it might even be that the team manager(s) would be asked to take the breakdown to a lower level again when these were being prepared.

One set of things we need to identify when product-based planning is what PRINCE2 calls **external products**, which already exist or are being created outside the scope of the plan. This can be relevant for stage plans and breakdown structures – for example if we need a product from a previous stage in order to create another one in the current stage, that first product is an external product or dependency for the current stage. But usually the most important context here is the project plan, because then an external product is something that we need on our project but we're not in control of its production. Especially if it's being produced by a parallel project, for example, the fact that it might be delivered to us late and hold up our own project is a potential risk. The confusion here is that we normally use the word *external* in the context of suppliers who are external to the host organisation, whereas here external products are external to the project. So, for example, a specialist product being produced for our project by an external supplier could never be an external product.

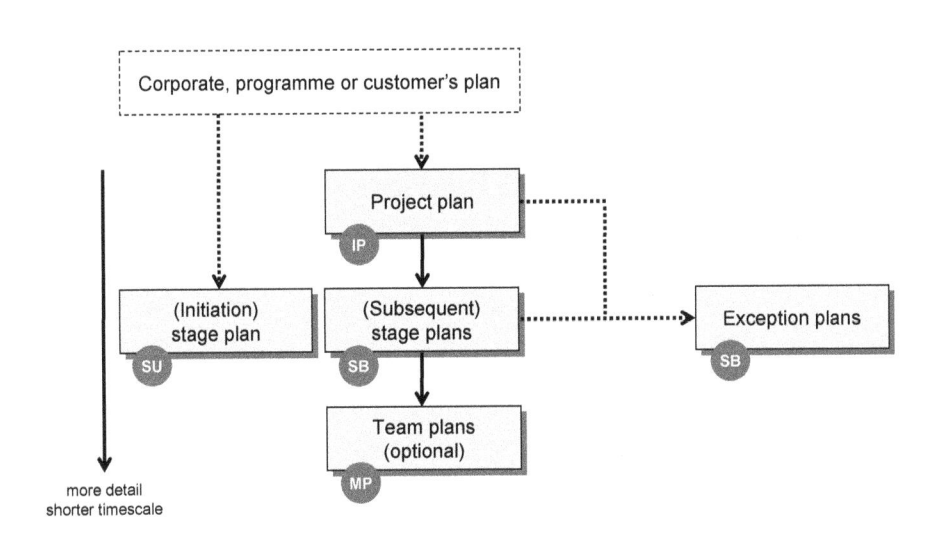

Figure 8: The 3 levels of plan

(based on Figure 9.1 Managing Successful Projects with PRINCE2® 2017 Edition)

9 THE PLANS THEME

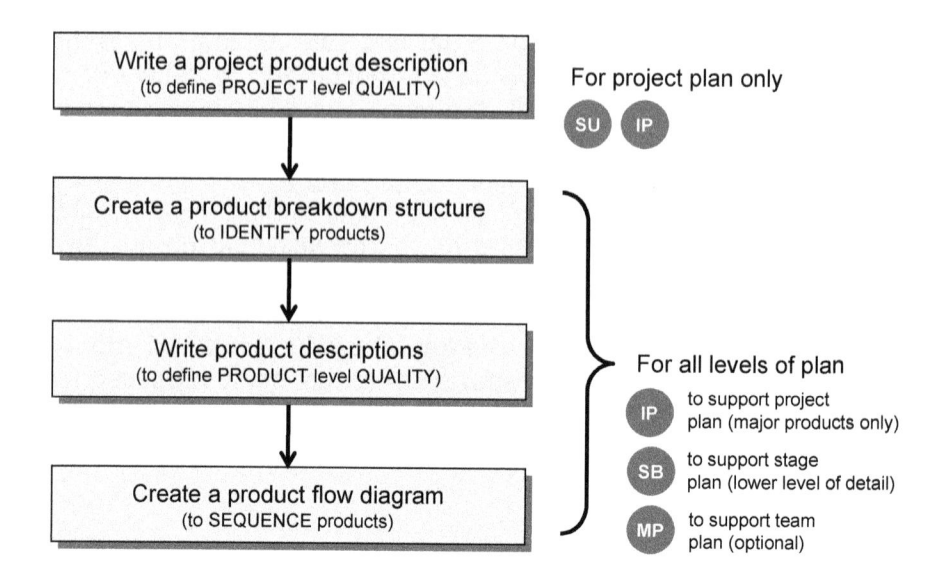

```
┌─────────────────────────────────────┐
│  Write a project product description │
│   (to define PROJECT level QUALITY)  │
└──────────────────┬──────────────────┘
                   │
                   ▼
┌─────────────────────────────────────┐
│  Create a product breakdown structure│
│         (to IDENTIFY products)       │
└──────────────────┬──────────────────┘
                   │
                   ▼
┌─────────────────────────────────────┐
│       Write product descriptions     │
│   (to define PRODUCT level QUALITY)  │
└──────────────────┬──────────────────┘
                   │
                   ▼
┌─────────────────────────────────────┐
│      Create a product flow diagram   │
│         (to SEQUENCE products)       │
└─────────────────────────────────────┘
```

For project plan only
- **SU**
- **IP**

For all levels of plan
- **IP** to support project plan (major products only)
- **SB** to support stage plan (lower level of detail)
- **MP** to support team plan (optional)

Figure 9: The product-based planning technique

(based on Figure 9.6 Managing Successful Projects with PRINCE2® 2017 Edition)

10 THE RISK THEME

Risk is all about <u>uncertainty</u>, which always exists on projects because they always introduce change of some sort. We can't just put our heads in the sand and ignore risk – it must be effectively <u>managed</u>, even on simple projects.

Different organisations will have different levels of **risk appetite**. Indeed some whole industries, like investment banking, have an inherently high appetite for risk, whereas retail banking at least ought to have a low risk appetite. We can also have different appetite for risk within one organisation – higher on a project that has the potential to deliver higher benefits, and vice versa.

In everyday life we tend to concentrate on **threats** that have potentially negative outcomes. But on a project we can also face **opportunities** that could have positive outcomes, and these too should be managed.

THE RISK MANAGEMENT PROCEDURE

The 5 steps in the recommended risk management procedure are summarised in Figure 10, and we'll look at each in turn. We will use a simple scenario as we go along to illustrate that the core of risk management on a project involves no more than putting some standard terminology on what we do automatically in our everyday lives, just using our common sense.

Step 1: Identify This is made up of 3 parts: the underlying **risk cause**, the **risk event(s)** it gives rise to, and the **risk effect** this creates. For example, every day we have to manage the risk of 'being late for work', which is the *effect*. If we're driving across town in the rush hour, then the *cause* is 'going by car', and the most obvious possible *event* is 'hitting traffic'.

Step 2: Assess When we assess a risk we're interested in quantifying 3 main things:

> ➢ The **probability**/likelihood of the risk *event*. In our scenario we might assess the probability of hitting traffic as 'high'.

- The **impact** or quantified effect of the risk on our 6 project objectives (as defined at the start of chapter 1) if it did occur. Most obviously we tend to concentrate on the potential impact on time, cost and quality when assessing a risk, and again we might just use a scale of high, medium or low.
- The **proximity** of the risk in terms of when it would occur. This is something we don't normally think about in our everyday lives because we tend to be managing imminent risks only, but in a project context they can be classified as, for example, *imminent*, *within the current stage*, *within the project* or *beyond the project*.

Step 3: Plan (our responses) Having assessed our risk, it's now time to decide how we want to respond to it, i.e. if there are any mitigating actions we might take. The following are the 5 different *types* of response to a *threat*:

- **Avoid** Such a response would completely eliminate the threat by taking its probability and/or impact down to zero. In our scenario, 'staying in a nearby hotel overnight' would pretty much eliminate the risk. So would 'getting the train', to the extent that it eliminates any traffic-related risk, but it introduces the *secondary risk* of the train itself being late or breaking down.
- **Reduce** By far the most common response to threats is to reduce their probability and/or impact but not down to zero, i.e. the *inherent risk* is reduced but there remains some *residual risk*. In our scenario 'leaving earlier' is an obvious example that would still leave some degree of threat.
- **Accept** Under certain circumstances (see below) it's perfectly reasonable to just accept a risk without taking any proactive action.
- **Transfer** This is an even more specific type of response where we transfer at least some of the financial risk to a third party. There is no obvious example for our simple scenario, but in our everyday lives we often do this by taking out insurance. In a project context inserting a penalty clause into a supplier contract e.g. for late delivery is a good example of

a transfer. Liquidated damages are even more effective.

> **Prepare contingent plans** Sometimes referred to as a 'fallback', this is a very specific type of response to a threat where we develop a 'plan B for if the risk happens'. The most obvious example of this in our scenario is to have alternative routes planned in advance that we can take if we hit traffic at some point on our journey.

The 3 main types of response to an *opportunity* are to **exploit** it (the equivalent of the avoid response to a threat, i.e. making sure it definitely happens), to **enhance** it (the equivalent of reduce) or, again, to just **accept** it. But less obviously we can sometimes transfer it or prepare contingent plans for it.

The other response to either a threat or an opportunity is to **share** it with a third party, especially when we don't know in advance which way the risk will go. For example, contracts with external suppliers often include a 'pain/gain' formula for sharing cost excesses or shortfalls compared to the plan.

Note that there are two parts to this step: first we must list our possible responses, and then we must select which of these are worth implementing. The way we make this choice is by weighing up the cost of the response against the combined probability and impact of the risk. So, for example, if all the possible responses were relatively high cost, yet the risk itself was only low probability and would only have a low impact even if it did occur, we'd just accept it and take no proactive action.

Step 4: Implement (and monitor) Rather than taking responsibility for all risks ourselves we can appoint separate **risk owners** for at least some of them. This will be the person best placed to implement any proactive responses we've agreed upon for that particular risk, and also to monitor it on an ongoing basis and to feed information back to us about whether it's improving, worsening or staying the same (in terms of its probability and/or impact).

The risk owner can be anyone, they don't necessarily have to be part of the management structure or even in the host organisation – for example we might ask someone in a regulatory body to do this for us if the risk involved forthcoming legislation. The risk owner can

10 THE RISK THEME

also delegate implementation of some or all of the risk responses to a **risk actionee**.

Step 5: Communicate This operates in parallel with all the other steps. But in particular each of the 4 progress reports – i.e. checkpoint, highlight, end stage and end project reports – has a section allowing the author to report the situation with major issues and risks upwards to the next level of management.

THE RISK BUDGET

Within our overall project budget we can 'ring fence' two separate budgets that are earmarked for specific purposes. One of these is a *risk budget* that would be used to pay for specific responses to identified risks and would also, ideally, contain an additional amount to cover as yet unidentified risks. This is because no matter how thoroughly we attempt to identify risks during initiation, new ones that we couldn't initially foresee will always crop up. The other is a *change budget*, which will be discussed in chapter 11.

THE RISK-RELATED MANAGEMENT PRODUCTS

The first of these is the **risk management approach**. As discussed in chapter 8 there are 5 of these management approaches, and this one defines the way we intend to manage risk on the project. In particular it describes the procedures, techniques and standards we'll use and the various responsibilities around them. Where this document is being prepared relatively formally, much of the input will come from an examination of the *risk context* in terms of the project's environment, complexity and so on, which forms part of the *identify* step.

The other is the **risk register**, in which all the information pertinent to each of the individual risks identified on the project is recorded.

PROBABILITY IMPACT GRIDS AND SUMMARY RISK PROFILES

These are more complex diagrams used when formally assessing and monitoring risks. They are explained in detail in chapter 10 of the official manual, but need not concern us here.

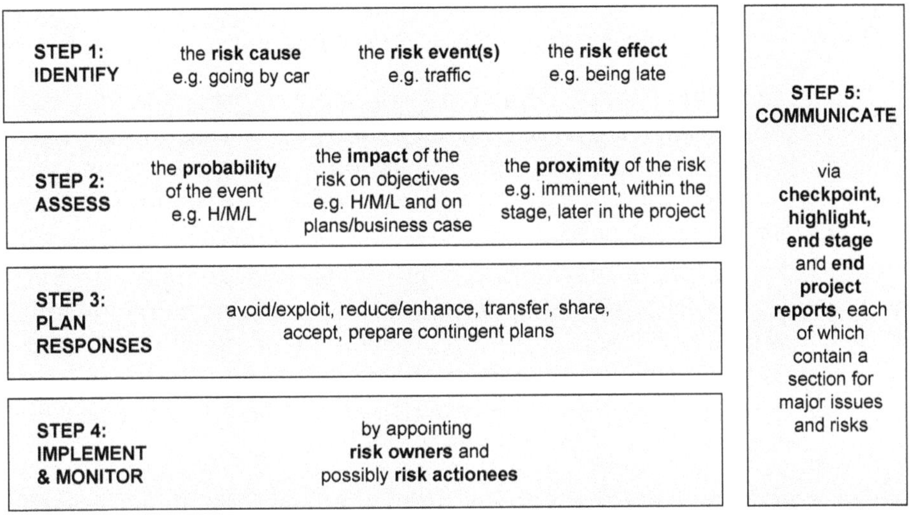

PRINCE2 MADE SIMPLE

STEP 1: **IDENTIFY**	the **risk cause** e.g. going by car	the **risk event(s)** e.g. traffic	the **risk effect** e.g. being late	**STEP 5:** **COMMUNICATE** via **checkpoint, highlight, end stage** and **end project reports**, each of which contain a section for major issues and risks
STEP 2: **ASSESS**	the **probability** of the event e.g. H/M/L	the **impact** of the risk on objectives e.g. H/M/L and on plans/business case	the **proximity** of the risk e.g. imminent, within the stage, later in the project	
STEP 3: **PLAN RESPONSES**		avoid/exploit, reduce/enhance, transfer, share, accept, prepare contingent plans		
STEP 4: **IMPLEMENT & MONITOR**		by appointing **risk owners** and possibly **risk actionees**		

Figure 10: The risk management procedure

11 THE CHANGE THEME

There are effectively 2 parts to the change theme:

- *Issue and change control:* it's important to grasp the fact that project issues are absolutely synonymous with the change theme, because these are the things that can cause changes to specifications and so on, and these changes need to be properly controlled.
- *Configuration management:* for those who struggle with this somewhat formal term you can think of this as being in large part about 'version control', which most people will have encountered in relation to documents, for example. Any change must always be discussed in relation to a *baseline* version to which the change is made.

So it should be clear that change and version control go very much hand in hand, and we'll look at each in turn.

ISSUE AND CHANGE CONTROL

If we don't have proper change control the project manager may tend to allow users to keep changing their minds about what they want, for example, which leads to diversion from agreed plans and 'scope creep'. The aim isn't to prevent changes, otherwise we'd be completely unresponsive to our stakeholders, but rather to make sure they're agreed by the relevant authority before they're made, and paid for out of the right pot of money.

Project Issues These can be about anything to do with the project, and can be raised by any stakeholder at any time. Users often raise issues, but for example team members working 'at the coalface' are obvious candidates to also raise issues while engaged in the creation of specialist products. There are 3 types of issue:

- **Request for change (RFC)** These encompass 'a changed or new or removed requirement, or a change in scope, or a suggestion for improvement'. Most obviously users tend to change their minds about what they want, and for example ask for a screen to have a dark grey background instead of light grey. But RFCs could come from a supplier too – for example a team member developing a specialist product might realise that they can do it cheaper and quicker and/or make it better. Or senior management might decide to authorise a major change in the scope of the project. All these would come under the heading of an RFC.

> **Off-specification** This is where a team manager/member indicates they can't fulfil all or part of a product description or more detailed specification document, typically because they've hit some sort of technical problem. In some cases it may be that the issue could be resolved if significantly more time and/or money, beyond the original tolerances, were allocated to that work package. This may or may not require a significant redesign of the product.

> **Problem/concern** This is any other issue not related to a specification. For example if a key resource that's in short supply falls ill or leaves the company, that's a major problem that will need our urgent attention.

Issues versus risks So that we're clear about the distinction between the two, an issue either has happened or definitely will happen, whereas a risk has some uncertainty attached to it. So let's say that as part of a marketing project we want to send a promotional calendar out to our existing and potential customers in the second week in December, but in late November we hear that postal workers are considering strike action that week. At this point it's a risk. Then in the first week in December we hear that talks have broken down and there will definitely be strike action the following week. Now the uncertainty has been removed, even though the event hasn't happened yet, so we'd transfer it out of the risk register and into the issue register.

Precisely because we may or may not authorise RFCs and therefore there's some uncertainty, some organisations prefer to have a separate 'change log' to handle them. PRINCE2 has no problem with this kind of tailored approach, but typically it regards RFCs as just another type of issue to be handled in the issue register.

The recommended change control procedure This is shown in

11 THE CHANGE THEME

Figure 11. A few aspects of this diagram are worth explaining in more detail:

> ➤ When we **capture** an issue we must first use our own discretion to decide whether to handle it *formally*, in which case we enter it in the **issue register** and optionally create a supporting **issue report** for more detailed analysis. Or if we decide to only handle it *informally* then we enter it in the **daily log** – which we can think of as the project manager's diary. If we're handling it formally we also have to record a) its *priority* (if it's an RFC), for which we might use a schema like MoSCoW (must have, should have, could have, won't have for now); and b) its *severity*, for example minor, significant, major or critical.

> ➤ When we **assess** an issue we're evaluating its *impact* on the 6 project objectives (as defined at the start of chapter 1), just as we do with a risk as we saw in chapter 10. We may also revise priority levels e.g. the apparently simple change of screen colour referred to earlier, which might have been initially rated by the user as 'must have', may now be downgraded if we find that it has a knock-on effect on a number of other modules or products. These revisions can be made in conjunction with ad-hoc advice from the project board.

> ➤ When we **propose** what we intend to do about a request for change this includes *evaluating* whether or not it's worth implementing by weighing up the (usually) extra cost and time against the estimated benefits – as if it were a mini business case.

> ➤ When we **decide** what to do with an issue, and particularly whether it needs to be escalated and if so to who, it depends on what type of issue it is. Cost and time tolerance is built into stage plans and work packages precisely to allow for estimating errors, off-specs and other problems that arise, so for these the pertinent question is: "Would the rectification of this issue(s) take me beyond my time and/or cost tolerance for the <u>stage</u>?" If it wouldn't we as the project manager can *take corrective action* and handle it ourselves, perhaps by revising the tolerances on the work package. If it would we

must escalate it to the project board via an exception report, as discussed at the end of chapter 5.

On the other hand *cost tolerance cannot be used to fund RFCs*. If we did use up our cost tolerance every time users changed their mind about what they wanted we'd have *scope creep* and, more important, we'd end up with no tolerance left to cope with off-specs and other problems. If a simple project was likely to only involve a few minor RFCs we might decide to bend this rule a little, but in most cases we should set up a delegated change authority with a separate change budget.

Note also that various aspects of the issue and change control procedure in Figure 11 are reflected in the activities in the controlling a stage process (see Figure 16).

The change authority and change budget The project board have responsibility for reviewing and approving RFCs and off-specs. However they can delegate this to a separate change authority. The project manager can act as their own change authority on a project – the control being that their change budget is 'ring-fenced' to pay for RFCs only – otherwise it will be taken on by those with delegated project assurance responsibilities.

Both the change and risk budgets may be set as a certain percentage of the overall project budget, usually based on project size and complexity, experience from previous projects and so on. They can also be allocated nonlinearly to individual stages, which is why the amounts set aside for them are recorded in the *budgets* heading in the relevant plan as well as under costs in the business case. The project board might also decide to place a limit on the amount that can be spent on any one change. Note that if we don't have enough change budget to pay for a change, especially e.g. for a major increase in scope, then we must ask for an increase in the overall project budget.

CONFIGURATION MANAGEMENT (VERSION CONTROL)

How often have you heard of construction projects where the architects keep changing the designs but the builders aren't informed? This is a result of poor change and version control. So

11 THE CHANGE THEME

what should happen? First, we should build the change procedures we want the architects to adopt into their contract, so that they're forced to obtain formal authorisation for any changes they want to make to the designs. Moreover, in estimating the impact of any proposed changes on time, cost and quality especially, the builders will have to be consulted, at least informally. If authorised any changes should then be built into a new version, which should be distributed to all relevant copyholders – including the builders – and we might even insist that all old copies should be returned/destroyed (if hard copy) or deleted (if electronic) to lessen the chance of them remaining in circulation. This is how change and version control dovetail together. The relevant procedures will be recorded in the **change control approach**, of which more shortly.

On larger projects we may need to use formal **configuration item records** – or, more simply, 'product records' – typically to document the versions of the different products, their change history and any linkages between them and so on. These are particularly important when large numbers of products that are going to change over time have multiple interrelationships with each other – so that if one is changed this has a knock-on impact on others. This information tends to be incorporated into a 'configuration database' made up of all the individual records for each product or configuration item. If used these records would be set up alongside and in addition to the product descriptions for the specialist products. In some organisations, for example those producing software or vehicles where components are usually required to work together, formal configuration management remains essential long after a development project is over as a key element of BAU.

As for management products, key documents that are regularly updated or sensitive to change will be subject to configuration/version control, such as the project plan, business case, project product description and individual product descriptions. By contrast logs and registers aren't subject to version control because they change _too_ regularly, so they tend to be held electronically with limited write access and unrestricted read access, so everyone is always looking at the most recent version. For full audit trail new entries in registers should be appended to the old entry rather than overwriting it.

THE CHANGE CONTROL APPROACH

Like the quality and risk management approaches discussed in chapters 8 and 10 respectively, this describes the procedures, techniques and standards that we'll use to achieve effective change and version control on the project, and the various responsibilities around them.

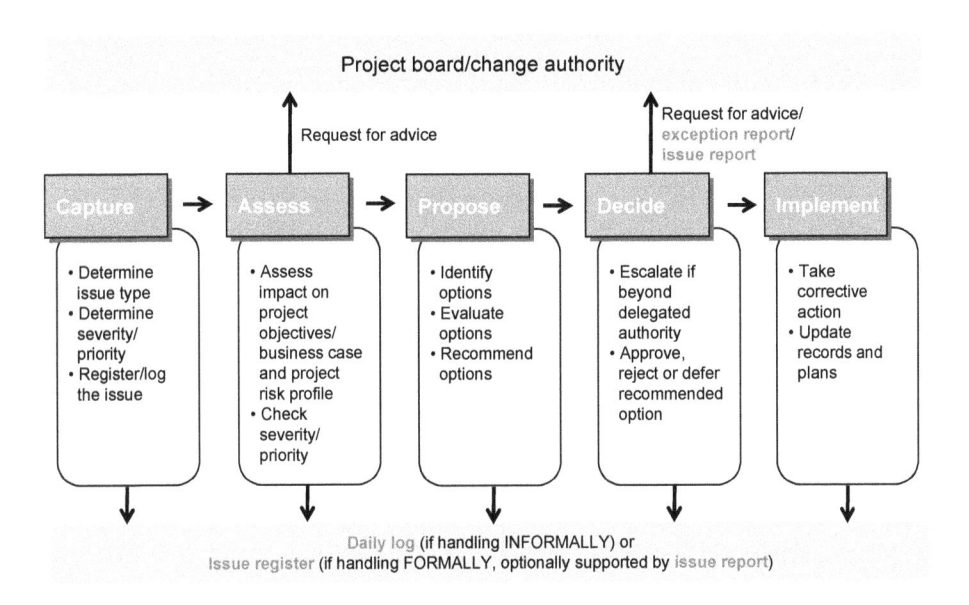

Capture	Assess	Propose	Decide	Implement
• Determine issue type • Determine severity/priority • Register/log the issue	• Assess impact on project objectives/business case and project risk profile • Check severity/priority	• Identify options • Evaluate options • Recommend options	• Escalate if beyond delegated authority • Approve, reject or defer recommended option	• Take corrective action • Update records and plans

Assess and Decide send requests upward to the Project board/change authority (Request for advice; Request for advice/exception report/issue report).

Output: Daily log (if handling INFORMALLY) or Issue register (if handling FORMALLY, optionally supported by issue report)

Figure 11: The issue and change control procedure

(based on Figure 11.1 Managing Successful Projects with PRINCE2® 2017 Edition)

12 THE PROGRESS THEME

There are effectively 3 parts to the progress theme:

- Tolerances and exceptions
- Time/event-driven controls
- Progress reports

Let us look at each in turn.

TOLERANCES AND EXCEPTIONS

We opened chapter 1 by discussing the 6 *project objectives*. Because these are parameters for which we can set measurable objectives, they're also the 6 things for which we can set measurable *tolerance limits*:

- **Cost and time tolerances** These are set for project plans, stage plans and work packages/team plans. They can be expressed in absolute terms (e.g. +£5k/-3k or +2 weeks/-1 week) or relative terms (e.g. +5%/-3%). Note that we can proactively set plus *zero* tolerances if a project absolutely must come in on time or budget. As for minus tolerances, we don't have to use these, but we will if we want to be informed that something is going to come in especially early or under budget – if, for example, we would then need to reallocate resources or unused budgets.

- **Quality tolerances** These have an entirely different context and are set for products – either in the project product description or the individual product descriptions. An example would be a product with a quality criterion that it must weigh 300g, but with an allowable tolerance around that of +/-10g.

  The three above tend to be the most commonly used tolerances, especially on smaller projects.

- **Benefits tolerance** This is any allowable tolerance around our forecast of benefits in the business case, and could come into play especially if we revised that forecast downwards during the project.

- **Scope tolerance** This would most obviously take the form of a MoSCoW-style prioritisation of *acceptance criteria* for the project product, but could also encompass some other allowable variation in the set of products to be delivered by a project/stage plan or work package.
- **Risk tolerance** Typically this involves the project board setting a risk tolerance limit based on a composite numerical scale derived from multiplying impact by probability, so that any risk with an impact-probability value greater than that limit would take us into exception (except in special circumstances).

Tolerances can also be traded off against each other. For example, if we had a project where it was vital to the business case that we met the end date, we'd almost certainly have a plus zero project time tolerance – but we might well be allowed a greater cost or quality tolerance than normal to help us to meet the all-important time target.

Setting/delegating tolerances Figure 12 shows the 4 levels of the management structure. On the left-hand side we're examining the setting/delegating of (primarily) time and cost tolerances. Remember that corporate, programme management or the customer are the sponsors of our project and provide us with our mandate. Especially if it's a formal mandate because our project is part of a programme, it's not hard to understand that the programme board will not only set the overall time and cost budgets for our project, but also any tolerances around those.

But let's now go to the other end of the scale, whereby a verbal mandate is passed down to us as project manager that has no mention of numbers in it at all. In this scenario we'd have to talk to users about what they really want, so we can prepare at least a draft of the project product description containing measurable *acceptance criteria*. Then, if we're not a subject matter expert, we'd have to get some kind of input from relevant suppliers about how much that might cost and how long it might take. We would then build these numbers into the outline business case, and possibly put some suggested project-level tolerances around them, all of which would form part of the project brief given to the project board at the end of

12 THE PROGRESS THEME

start up. But would it be up to the project board to ratify these numbers at this point? No. They would have to go back to corporate management or the customer who, as the project sponsors, not only provide the budget but also set any project-level tolerances for time and cost.

Once we understand that corporate, programme management or the customer set the overall project-level tolerances they want the project board to work within, it's easy to see in Figure 12 that the project board then set the stage-level tolerances they expect the project manager to work within, who in turn sets the work package/team plan tolerances they expect the team manager(s) to work within.

Monitoring tolerances Coming back up the right-hand side of Figure 12, the team manager(s) reports on the work package progress and tolerance situation via checkpoint reports, the project manager reports on the stage progress and tolerance situation via highlight reports, and the project board reports on the overall project progress and tolerance situation most obviously via end stage reports, although corporate, programme management or the customer could ask to be kept more regularly informed via highlight reports too.

Escalating issues/exceptions Staying on right-hand side of Figure 12, if the team manager(s) is forecasting that they can't complete a work package within time and/or cost tolerance, they must escalate this to the project manager via a project issue (not an exception at this level). If this doesn't threaten the stage tolerances the project manager can take their own *corrective action*. On the other hand, as we saw at the end of chapter 5, if the stage tolerances are threatened for whatever reason the project manager must escalate this to the project board via an exception report. If this doesn't threaten the overall project tolerances the project board can ask for an exception plan to replace the current stage plan, and we carry on. On the other hand if the overall project tolerances are threatened for whatever reason the project board must escalate this to corporate, programme management or the customer, again via an exception report prepared by the project manager. If this doesn't threaten the business case they will probably ask the project board to carry on, by providing the project with additional funds and/or time as appropriate, and the project board in turn will ask for an exception plan to replace the overall project plan. However if for whatever reason the business case is threatened then if possible corporate, programme management or the customer will authorise a major change in scope to rescue the project, usually a de-scoping, otherwise they'll be forced to authorise a premature close.

Note that all decisions about additional budget, major changes in scope or premature closure are actually made by corporate, programme management or the customer. But because the project board act as their interface to the project these decisions are filtered back down to us via the project board in the directing a project process.

It is also vital to remember that we go into exception when we *forecast* that we can't complete the stage or project within cost and/or time tolerance. We don't wait until we've got to the end and then ask for more money or time. This is because we need to be 'proactive not reactive' about exceptions. Most obviously this is so that if, for example, there's a major exception situation that threatens the business case, we're giving corporate, programme management or the customer (via the project board) the chance to pull the plug on the basis that we might have spent £x million, but we're not going to waste another £y million on a project that's no longer worth doing.

TIME VERSUS EVENT-DRIVEN CONTROLS

There are only 2 purely *time-driven* controls in PRINCE2: these are checkpoint reports and highlight reports, because they're produced at predefined, periodic intervals – typically weekly and monthly, but this is by no means a fixed rule. In fact the formality and regularity of highlight and checkpoint reports is determined by the risk and complexity of the stage and work package respectively. Remember we shouldn't be having progress-only meetings with the project board when we produce highlight reports within a stage. But if a work package was particularly high risk – involving, for example, a new external supplier, and the development of complex, state-of-the-art products – a checkpoint report could take the form of a formal meeting; this might be weekly, but at a particularly risky point in a

12 THE PROGRESS THEME

project it could even be daily. By contrast if we're dealing directly with one or two team members on a low risk, simple work package, we might keep it very informal and just ask them to phone us once a week to let us know how they're getting on – making notes of the call in our daily log. Checkpoints are good examples of tailoring PRINCE2 using common sense.

All other controls in PRINCE2 are *event-driven* in that they're triggered by non-periodic events. The most obvious events that trigger various controls to come into play are, first, stage boundaries and, second, the production of an exception report.

PROGRESS REPORTS

There are 4 progress reports in PRINCE2:

- ➢ **Checkpoint reports**: produced by team manager(s) when managing product delivery, these typically contain a summary of products worked on/completed since the last checkpoint, a forecast of what's planned for the next period, a tabular summary of the work package progress and tolerance situation in terms of actual and forecast time and cost, and an update on any major issues and risks.
- ➢ **Highlight reports**: produced by the project manager when controlling a stage, these typically contain a summary of products and work packages issued/worked on/completed since the last highlight, a forecast of what's planned for the next period, a tabular summary of the stage and project progress and tolerance situation in terms of actual and forecast time and cost, and an update on any major issues and risks.
- ➢ **End stage reports**: produced by the project manager when managing a stage boundary, these typically contain a tabular summary of the stage and project progress and tolerance situation in terms of actual and forecast time and cost, and an update on any major issues and risks.
- ➢ **End project report**: produced by the project manager when closing a project, these typically contain a tabular summary of the project performance against objectives for time, cost

etc., formal acceptance records, a lessons report and details of any issues and risks that will need to be addressed post project.

SETTING & DELEGATING **MONITORING & ESCALATING**

Sponsors/mandate	Corporate, programme management or the customer	Make decisions about extra budget/ premature close

Project tolerances ↓	End stage (highlight) reports	Project progress/exceptions ↑	Exception report

Project board

Stage tolerances ↓	Highlight reports	Stage progress/exceptions ↑	Exception report

Project manager

Work package tolerances ↓	Checkpoint reports	Work package progress/issues ↑	Project issue

Team manager

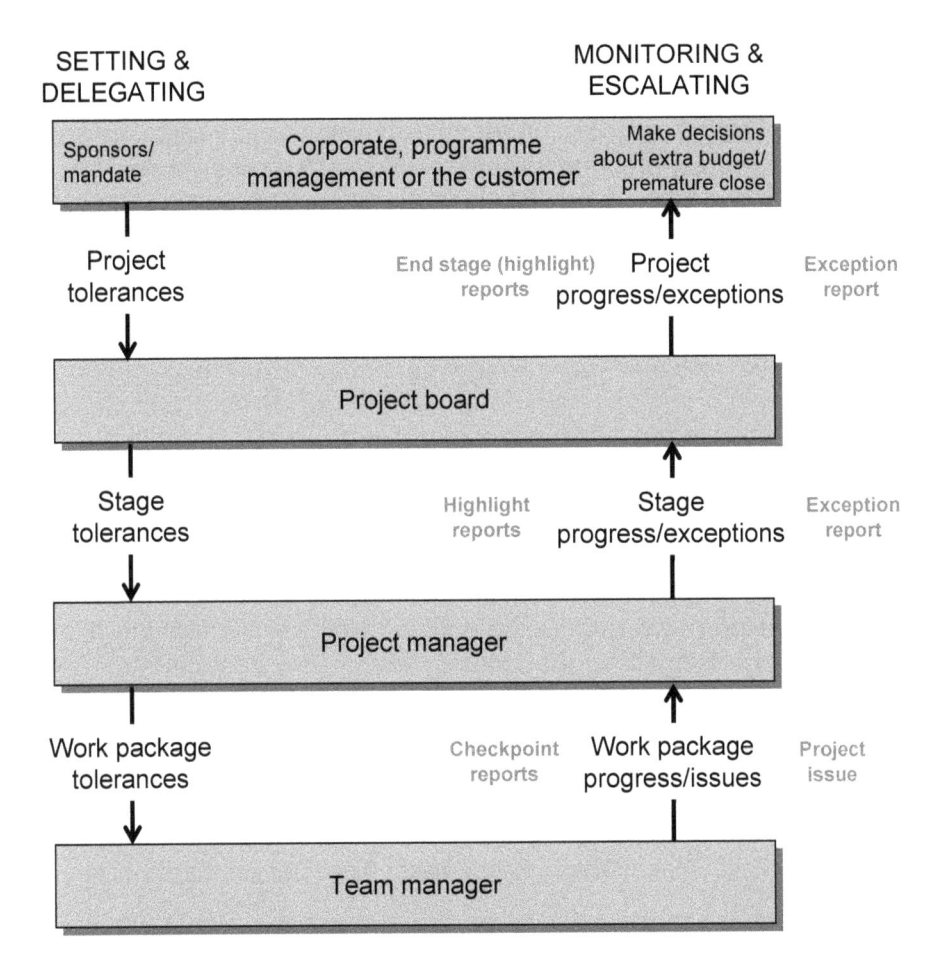

Figure 12: Setting/monitoring tolerances and escalating exceptions

(based on Figure 12.1 Managing Successful Projects with PRINCE2® 2017 Edition)

13 STARTING UP A PROJECT PROCESS

[Note that directing a project doesn't have its own chapter in what follows, but is covered within all the other process chapters. Note also that in the detailed process diagrams in this and the following chapters, the lighter grey rectangles represent the individual activities within the process under consideration, and the darker grey rectangles represent other processes with which it's interacting – while within the latter the individual activities are shown in a smaller font where relevant. Management products or other important documents or parts thereof that are likely to be written down in some fashion are shown in grey and bolded, while themes are shown in grey ovals and abbreviated as per the list in chapter 2. Remember also that each process represents only a *checklist of recommended activities*, which don't all have to be carried out formally and can be tailored to suit the individual project.]

As we saw in chapter 4 start up is a separate process from initiation, and indeed regarded as *pre-project* work, because it's something of a 'finger in the air' exercise – in which the aim is to do the minimum necessary to see if it's worthwhile to undertake a full initiation of the project. We don't want to waste time and money performing a full initiation only to establish that actually the 'quality-cost-time triangle' is nowhere near equilibrium (see chapter 1), and/or the business case isn't viable. So we produce only a draft project product description and an outline business case, in which the estimates of cost and time for the project as a whole aren't based on a proper project plan. If start up is used properly and estimates of benefits aren't overoptimistic, many projects won't even get off the ground.

If we now refer to Figure 13, corporate, programme management or the customer trigger start up by issuing a **project mandate**. As a minimum this identifies the executive on the project board and provides some sort of terms of reference that tell us what the objective of the project is. But remember this could be just a verbal mandate containing no other details. In practice, also, the ideas for projects can come from lower down in the host organisation – but if so someone from this top layer of management must still decide which ideas are prioritised for investigation.

APPOINT THE EXECUTIVE AND THE PROJECT MANAGER

The project mandate appoints the executive, who in turn appoints the project manager. The mandate is then effectively handed to the project manager so they can get on with the other activities in this process. The **daily log**, or project manager's diary, is set up at this point.

CAPTURE PREVIOUS LESSONS

This activity encourages us to proactively work through previous projects' lessons reports to seek out both good and bad lessons that might apply on our project, and to capture them in the **lessons log**. In reality this might take some time, though, so during start up we might content ourselves with discussions with other project managers who've undertaken similar projects, and complete the more detailed examination of lessons reports or databases during initiation.

DESIGN AND APPOINT THE PROJECT MANAGEMENT TEAM

Even though like the mandate it's not defined as an official management product, the **project management team structure** is part of the project brief and the PID, and will need to be documented in some way. Again we might only design a skeleton team during start up, and flesh it out properly during initiation.

In particular on a larger project we might write formal **role descriptions** defining the detailed responsibilities allocated to each person appointed to any given role, perhaps based on the proformas provided in Appendix C of the official manual – although again this might be delayed until initiation. These can be especially useful if we face the prospect of, for example, project board members who may be highly enthusiastic at the outset and then more reluctant to make their resources available later on.

PREPARE THE OUTLINE BUSINESS CASE

We have already discussed the **outline business case** at length in chapter 1 and elsewhere, along with the **project product description** that should be drafted at the same time. Note that the

13 STARTING UP A PROJECT PROCESS

risk theme has been associated with this activity in Figure 13, to reflect the fact that we must at least identify major risks to the potential project in the outline business case, even if during start up we don't yet have a risk register. These will initially be recorded in the daily log.

SELECT THE PROJECT APPROACH & ASSEMBLE THE BRIEF

The general **project approach** document examines our 'options in terms of products and resources'. For example, will we be buying our project product off the shelf or developing it from scratch? Will we be using in-house or external resources or a combination of the two? Will we or our suppliers be adopting an agile approach to product development? It is important to undertake this formally on larger projects where the overall approach to the project may not be obvious and various options need to be considered.

The **project brief** is made up of a general definition section, the outline business case, the project product description, the project management team structure and role descriptions, and the project approach. On a larger project it will be a composite of a number of individual and separate documents.

PLAN THE INITIATION STAGE

In order for the project board to *authorise initiation* they will also need an **initiation stage plan** – although again this might be delayed until we've got unofficial confirmation that the outline business case looks good and initiation will go ahead. If this is formally confirmed the end of start up marks the 'official start' of the project. This is also when the directing a project process officially commences.

[The various outputs of start up are also shown in the completed Summary Diagram in Figure 19. Note that although to keep it simple start up is shown as one of the project manager's processes, the initial activity of appointing the executive and authorising the latter's choice of project manager is undertaken by corporate, programme management or the customer, as shown in Figure 13.]

PRINCE2 MADE SIMPLE

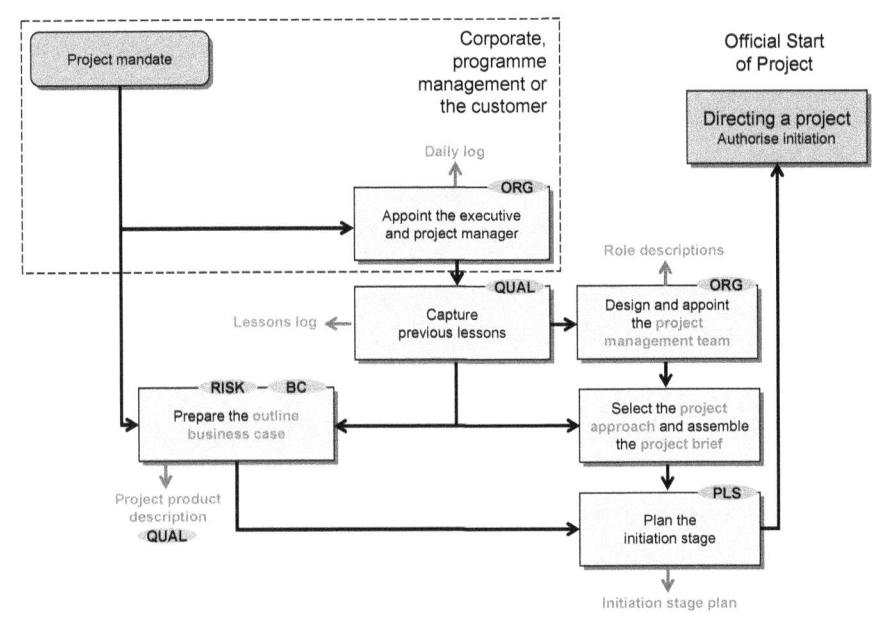

Figure 13: Starting up a project process

(based on Figure 14.1 Managing Successful Projects with PRINCE2® 2017 Edition)

14 INITIATING A PROJECT PROCESS

If we refer to Figure 14 we can see that initiation, which is also the first stage in a PRINCE2 project, is triggered when the project board *authorise initiation*.

AGREE THE TAILORING REQUIREMENTS

The first activity is to decide how PRINCE2 is going to be tailored for this particular project, including any departures from the norm for the organisation. On larger projects this will be documented under its own heading in the PID. Tailoring will be discussed further in chapter 20.

PREPARE THE FIRST FOUR MANAGEMENT APPROACHES

The next four activities involve preparing more detailed approach documents covering **risk**, **quality**, **change** and **communication**, all of which have already been discussed in chapters 10, 8, 11 and 7 respectively. At the same time as preparing the first three of these we'll set up the **risk, quality** and **issue registers**.

SET UP THE PROJECT CONTROLS

The purpose of this activity is to actually set up the controls that will have been decided on, at least partially in the above approach documents. There is also a separate heading in the PID in which on larger projects we can separately document the various **project controls** that have been agreed.

CREATE THE PROJECT PLAN

The **project plan** is one of the main 'drivers' of the project and something we didn't attempt to create during start up because it would potentially take too long for a 'finger in the air' exercise. But now we are creating it we'll use the main elements of the *product-based planning technique* for the first time, particularly to produce a **product breakdown structure** identifying the major specialist products of the project, then creating **product descriptions** for each one (see chapters 8 and 9). If we're undertaking formal configuration management we'll also be creating a first set of **configuration item**

records for them at this point (see chapter 11). In addition this is where we'll update and refine the **project product description**, another of our main 'drivers'.

Note that the **project approach** (see chapter 13) acts as a key input to this activity because it may have a significant impact on our plan.

PREPARE THE BENEFITS MANAGEMENT APPROACH

Having created the project plan for the first time we can now refine the outline **business case** produced in start up, not just by firming up on the estimates of cost and time but also, for example, of benefits. This completes our final main 'driver'. In addition this is the point at which we create the **benefits management approach** (see chapter 6).

ASSEMBLE THE PROJECT INITIATION DOCUMENTATION

The **PID** is made up of a general definition section, the business case, the project management team structure and role descriptions, and the project approach (all refined and updated from the brief); 4 of the detailed approach documents (note the benefits management approach is kept separate from the PID because it will be required post project); the project plan (which now includes a refined project product description); and possibly sections on project controls and on how PRINCE2 will be tailored to this project. On a larger project it will be a composite of a number of individual and separate documents.

MANAGING A STAGE BOUNDARY

In chapter 5 we saw that at the end of initiation we'll also switch into a 'mini version' of managing a stage boundary, mainly be so we can produce a **next stage plan** and on a larger project an **end stage report** too – meaning we won't be undertaking the 'housekeeping' with registers, logs and so on that typically forms part of this process.

It is not impossible that we might lose *business justification* during initiation, for example if we have to heavily revise our estimates of costs, timescales or benefits, so an *end stage assessment*

14 INITIATING A PROJECT PROCESS

associated with a stage boundary will be undertaken – although premature closure at this point should be relatively rare. Normally the business case will still be viable so the project board will take the decision to *authorise the project* – remember this should be interpreted that they're authorising the *specialist work* to begin – at the same time as they *authorise our stage plan* for stage 2.

[The various outputs of initiation are also shown in the completed Summary Diagram in Figure 19.]

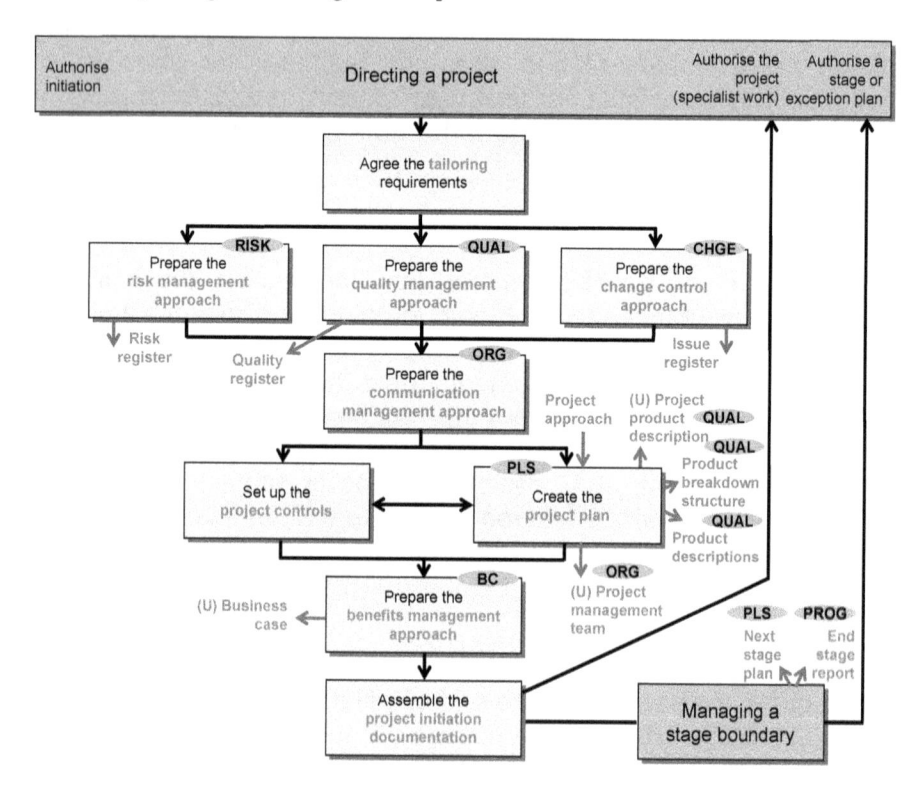

Figure 14: Initiating a project process

(based on Figure 16.1 Managing Successful Projects with PRINCE2® 2017 Edition)

The next process to be triggered by the above decisions will be controlling a stage, because we need to start controlling the first specialist stage. But in doing so our first order of business is to negotiate work packages with the team manager(s), so in the next chapter it makes sense for us to examine the managing product delivery process first.

15 MANAGING PRODUCT DELIVERY PROCESS

If we refer to Figure 15 we can see that managing product delivery is triggered when the project manager *authorises a work package* in the controlling a stage process, which we'll consider in the next chapter. It may be worth referring back to the Line Diagram in Figure 3 at this point, as well as to the Summary Diagram in Figure 19, before we look first at what's happening from the perspective of the team manager(s).

ACCEPT A WORK PACKAGE

Ideally the acceptance of a **work package** should be a negotiation between the project and team manager(s) – because the latter will respond to this far better than blanket imposition. As a minimum it will need to make reference to the cost and time allocated to this work, any tolerances around these, and to the product descriptions of the specialist products to be procured or created. For a larger piece of work – particularly one being undertaken by, for example, a new external supplier and involving complex and state-of-the-art products – it may also refer to procedures for reporting, change and version control, raising issues and risks, and other more detailed techniques, interfaces or constraints to be adhered to.

Optionally the team manager(s) may be asked to create a **team plan** at this point to show how the work package will be produced.

EXECUTE A WORK PACKAGE

The team manager(s) then have to get the actual work done by managing the team members 'at the coalface'. While this is going on they will produce regular **checkpoint reports** for us (see chapter 12), typically but not necessarily weekly. Remember that the formality of these will vary according to the risk and complexity of the work package.

Remember too that the team manager(s) are responsible for *quality control* (see chapter 8). In practice, all too often when a product is completed the team manager hands it over to the project manager to get it checked. The reviewers find that it has various problems and hand it back to the project manager who in turn communicates all this to the team manager and instructs them to carry out the necessary reworking. PRINCE2 suggests there's no need for the project manager to be continually stuck in the middle of this process, and that it's much more streamlined if the team manager liaises directly with the reviewers to make sure all the products in the work package end up being approved before it's handed back.

Note that there's no lack of independence here – the same independent reviewers/users are still checking the products, it's just that the team manager rather than the project manager is responsible for getting sign off. This also means that the team manager(s) are responsible for making sure the **quality register** or 'diary of quality checks' is updated with the results of the quality checks (again see chapter 8).

DELIVER A WORK PACKAGE

All the foregoing means that, when the team manager(s) hand the completed work package back to the project manager, it's 'fully quality checked'.

[The various outputs of managing product delivery are also shown in the completed Summary Diagram in Figure 19.]

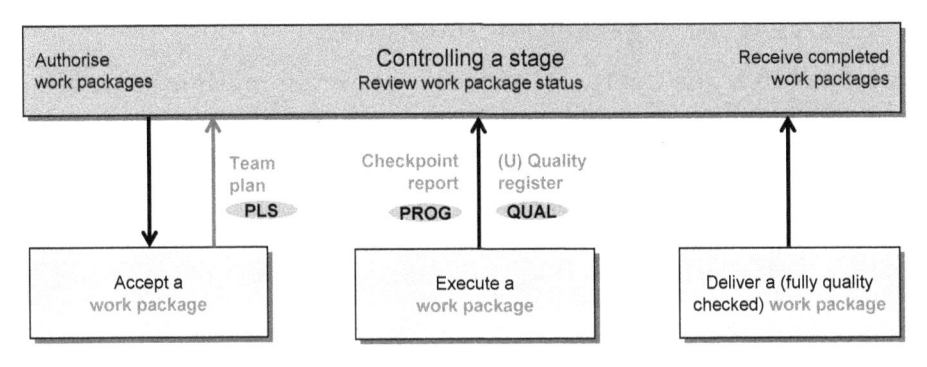

Figure 15: Managing product delivery process

(based on Figure 18.1 Managing Successful Projects with PRINCE2® 2017 Edition)

16 CONTROLLING A STAGE PROCESS

If we refer to Figure 16 we can see that controlling a stage is triggered when the project board *authorise a stage (or exception) plan* at the boundary between stages.

AUTHORISE WORK PACKAGES

As we saw at the end of chapter 14 the first order of business is to authorise **work packages**, which mirrors the team manager(s) *accepting* them. There might only be one in the stage, or there might be many. If there are many these might all need to be started at the same time, or one after the other, or a combination of the two.

REVIEW WORK PACKAGE STATUS

This mirrors the team manager(s) *executing a work package*, during which they produce regular **checkpoint reports** for us so we can check on their progress. Remember these could take the form of a meeting, a written report or just a phone call. Using these we'll update the detailed **stage plan** with actual figures for cost and time – typically, although not necessarily, with the same frequency with which we receive checkpoints, e.g. weekly.

RECEIVE COMPLETED WORK PACKAGES

This mirrors the team manager(s) *delivering a work package*. Remember that when it's complete it will already be fully quality checked, in that all the products will have been approved.

CAPTURE AND ASSESS ISSUES AND RISKS

Apart from concentrating on what's happening with the work package(s), we also need to manage our issues and risks on an ongoing basis. This involves monitoring those we've already identified, and capturing new ones, via the **risk register** and **issue register**. If we're handling an issue formally in the register we may also need to create an **issue report**, while if we choose to handle it informally we'll simply enter it in the **daily log** (see chapter 11).

REVIEW THE STAGE STATUS

This broader review that we undertake covers not only what's happening with the work packages, but everything relating to issues and risks and so on as well. From a project manager's perspective this is probably the most important activity of all, which is why we might refer to it as the 'engine room'. Ideally we should take time out typically once a week (but dependant on the size and complexity of the project) to rise above all the details we tend to get bogged down in during meetings, phone calls and so on – and to take the 'helicopter view' or see the 'big picture'. We want to make sure we don't miss the 'elephant in the room'. This is a hugely important part of being a professional project manager, and we should always be prepared to defend the fact that we might have put a 'do-not-disturb sign' on the door and switch our phone off for an hour or two a week. As part of this activity we might also ask for ad hoc advice about issues or risks from the project board, for example by passing an important **issue report** to them.

REPORT HIGHLIGHTS

Typically but not necessarily once a month we'll prepare a **highlight report** for the project board (see chapter 12). This will almost certainly be documented, but remember it shouldn't involve having a progress-only meeting with them.

TAKE CORRECTIVE ACTION

While we're periodically *reviewing the stage status* we have to ask ourselves a very important question: "Am I forecasting that I can complete this stage within cost and/or time tolerance?" If our answer is "Yes" we might still have one or more work packages that a team manager is forecasting can't be completed within tolerance. As long as this doesn't threaten the stage tolerances, as the project manager we can take our own *corrective action* and rework the affected work package(s) with the team manager(s).

ESCALATE ISSUES AND RISKS

On the other hand if our answer to the above question is "No" then we must escalate the situation to the project board via an **exception**

16 CONTROLLING A STAGE PROCESS

report. This might be because we're generally reviewing our progress and we can see that cost or time is slipping too much, or it could be related to a specific risk or issue and therefore the risk or change themes. Note that this activity, along with *take corrective action* and *capture and examine issues*, can be seen on the change control diagram in Figure 11.

MANAGING A STAGE BOUNDARY AND CLOSING A PROJECT

When we're getting near the end of the stage we'll make the decision for ourselves, while *reviewing the stage status*, that we need to switch over to the managing a stage boundary process. Or, if it's the end of the final stage, to the closing a project process.

[The various outputs of controlling a stage are also shown in the completed Summary Diagram in Figure 19.]

PRINCE2 MADE SIMPLE

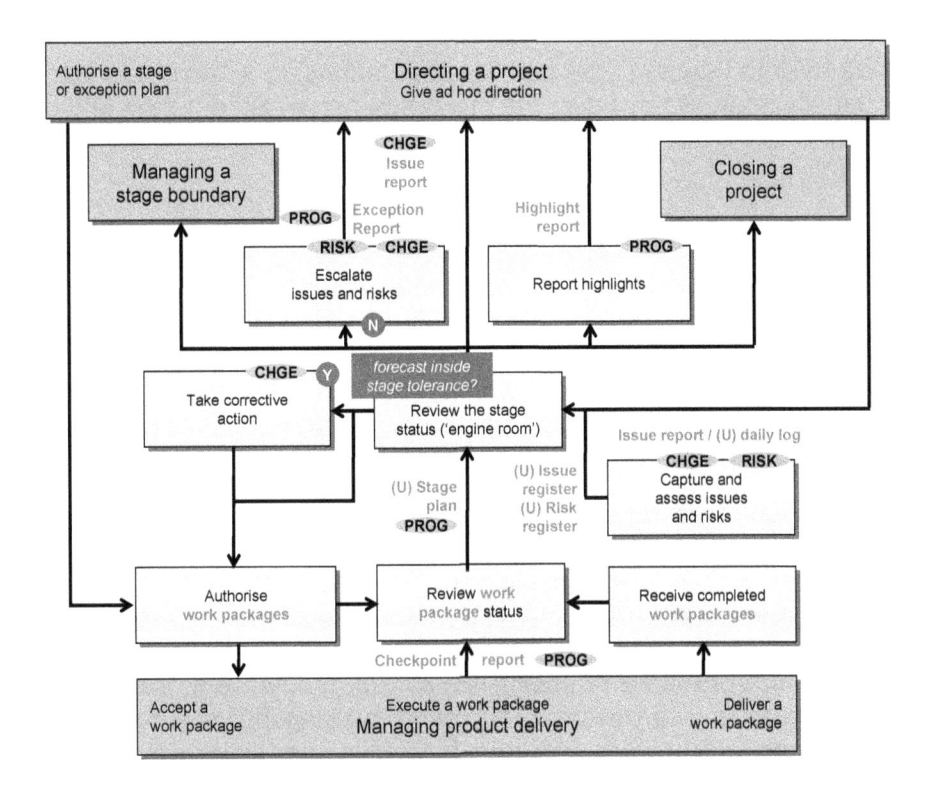

Figure 16: Controlling a stage process

(based on Figure 17.1 Managing Successful Projects with PRINCE2® 2017 Edition)

17 MANAGING A STAGE BOUNDARY PROCESS

If we refer to Figure 17 we can see that managing a stage boundary is usually triggered by us as the project manager, either at the end of *initiation* or towards the end of a specialist stage that we've been *controlling*.

PLAN THE NEXT STAGE

Our first order of business is to create a **next stage plan**. While doing this we may be using the *product-based planning technique* to revisit the **product breakdown structure** and identify new, lower-level, specialist products, in which case we'll need to write new **product descriptions** (see chapters 8 and 9) and possibly **configuration item records** (see chapter 11).

We will also enter the planned quality checks for the products of the next stage into the **quality register** (see chapter 8). In addition we may want to put new people into various roles in the **project management team**. Most obviously these would be the supply-side roles of team manager(s) and senior supplier(s) because we're moving from one area of specialist work to another, e.g. from using architects to create designs and plans to using builders to do the actual construction work. But it could involve user-side roles too e.g. in a phased implementation that's moving from one user location to another.

UPDATE THE PROJECT PLAN

Next we'll update the **project plan**, both with our actual progress to date in terms of time and cost for the stage just completed, and with a revised forecast forwards for the project as a whole – based on having a detailed stage plan for the next stage for the first time, and on any revised forecasts for the stages after that.

UPDATE THE BUSINESS CASE

Having done that we can now revise the time and cost forecasts in the **business case**, and possibly the estimates of expected benefits too. We may also update the **benefits management approach** at this point, particularly if any in-project benefits have already been realised (see chapter 6).

Note that these last two activities involve updating two of the main 'drivers' of any project, and since they both form part of the PID this too is automatically being updated. Note too that all of the first three activities also require us to update the **issue register** and the **risk register**. On a reasonable-sized project we should be doing this formally anyway while we're controlling a stage but, if not, as a minimum we must make sure they're up-to-date and fully reviewed at each stage boundary, as a form of 'housekeeping'.

REPORT STAGE END

The main output produced for the project board at a stage boundary is an **end stage report**, which summarises our progress (see chapter 12) and in particular any significant updates to the project plan and/or business case. At this point we'll also do some further housekeeping by updating our **lessons log** with anything that's been done particularly well or badly in the stage just completed (on a larger project we might even produce a formal **lessons report** at the end of a stage).

END STAGE ASSESSMENT

The project board will undertake this using the end stage report and next stage plan. Remember what they're primarily interested in is whether the fundamental principle of *continued business justification* still holds good. If it does they'll *authorise our stage plan*. However if things have been slipping away from us, or possibly because we deliberately placed this stage boundary at a major risk point (see chapter 9), they may need to consider a major *change in scope* to maintain a viable business case (usually but not necessarily a reduction), or even a *premature close*. Remember that either of the latter would have to be authorised by corporate, programme management or the customer, but the decision would be filtered down to us via the project board (again see chapter 12).

PRODUCE AN EXCEPTION PLAN

While we're controlling a stage, if we forecast that we can't complete

17 MANAGING A STAGE BOUNDARY PROCESS

the stage or project within time and/or cost tolerance, we escalate this to the project board via an **exception report** (see chapter 16). Provided the business case isn't threatened, and possibly with extra budget and/or time allocated by corporate, programme management or the customer if it's a project-level exception, the project board will then ask us to produce an **exception plan**. This will force an 'exceptional stage boundary' that we didn't originally expect to have, switching us into this process and this activity (see chapters 5 and 12). The new plan will replace either the existing stage or project plan, whichever is in exception. We may then have to carry out some of the other activities in the process too, although this will probably only be a 'mini version' that doesn't include all of the normal 'housekeeping' of registers, logs and so on.

[The various outputs of managing a stage boundary are also shown in the completed Summary Diagram in Figure 19.]

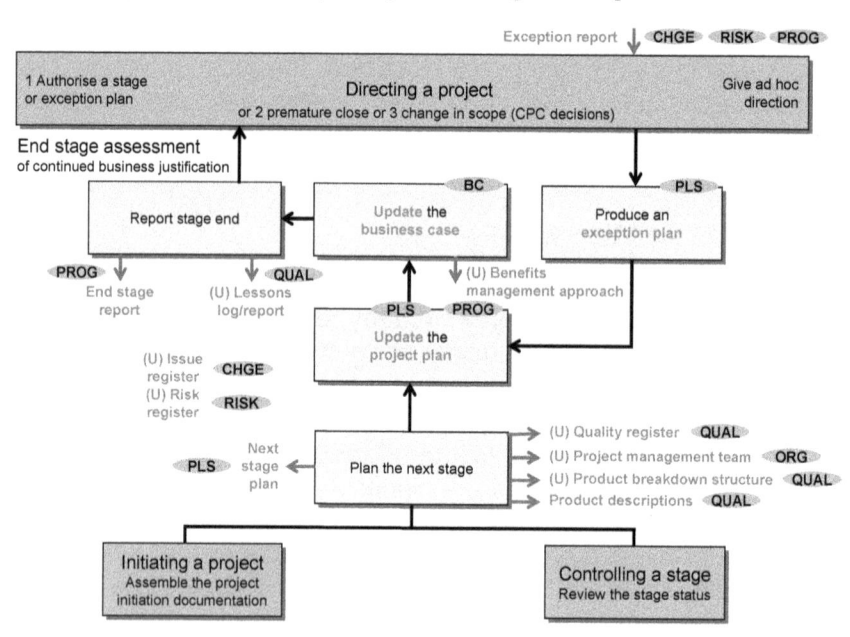

Figure 17: Managing a stage boundary process

(based on Figure 19.1 Managing Successful Projects with PRINCE2® 2017 Edition)

18 CLOSING A PROJECT PROCESS

When we're closing a project we're interested in evaluating two very different things:

- First, has the project been a success? To answer this we can always ask the same question, irrespective of the nature of the project: "Have we delivered quality products (although remember these might be intangible e.g. new/revised processes or systems) within the time and cost allocated to us?" If the answer is "Yes" we've run a successful project.
- Second, is the project product or final output a success? This is evaluated in two ways. While we're closing the project we'll need to make sure it's acceptable to the users and/or operational team who are going to use and/or maintain it. But, from a broader perspective, will it be a success in its operational life and deliver the benefits expected of it? This can normally only be answered after the project is over by conducting a series of *benefits reviews* (see chapter 6). If it does fall short, is it our fault as the project manager? Undoubtedly not. Remember it's up to the senior user(s) to buy into the forecasts of anticipated benefits used to justify the project in the business case, so it's them who should be held to account if these aren't realised.

Often in practice we find that closing a project isn't well performed, especially in that it's often unclear when the development project is complete and we've switched into an operational/live/BAU environment. PRINCE2 has no magic answer to exactly when this point is on any given project, but it does rightly insist that such a point – which, remember, should dovetail with our estimates of timescales in the project plan and business case – must be defined and clearly understood by all stakeholders. We should also be clear that when we close the project the whole project management team will be disbanded, and the only people left will be corporate, programme management or the customer – who will need to make sure the benefits reviews are carried out according to the plan – and nominally the senior user(s) who will be held to account if benefits fall short. This also means that if any external suppliers in the

18 CLOSING A PROJECT PROCESS

development team will have an ongoing maintenance role, the nature of the host organisation's contract with them needs to change.

Turning now to Figure 18, closing a project is triggered in one of two ways:

PREPARE PLANNED CLOSURE

A *planned closure* is triggered by us as the project manager while we're controlling (the final) stage. At this point we'll update the **project plan** with final actuals for time and cost.

PREPARE PREMATURE CLOSURE

A *premature close* is triggered by the project board when *giving ad-hoc direction* or when *authorising a stage or exception plan*, although the actual decision will have been taken by corporate, programme management or the customer (see chapter 12). In this case we'll undertake a somewhat cut-down version of closing a project, which is unlikely to involve handing over any project products or final outputs because we're unlikely to have completed them. Nevertheless we should make sure documentation is properly updated and so on, especially if there's a possibility we may pick the project up again at some point in the future.

HAND OVER PRODUCTS

The main priority here is to make sure we obtain formal, written **acceptance records** from the users and/or operational team who will use and/or maintain the project product(s) after the project is over.

We will also need to identify any open issues in the issue register, e.g. RFCs that were previously turned down but can be considered for a new version 2 once version 1 has gone live, or any open risks in the risk register that could affect the project product in its operational life. These will be transferred into a **follow-on action recommendations** document to be passed to the users and/or operational team, so the issue and risk registers can be closed down.

In addition we'll undertake a final update of the **benefits management approach.** The project board will ultimately pass it on to corporate, programme management or the customer to ensure any planned post-project benefits reviews get carried out.

EVALUATE THE PROJECT

The main way in which the project board and corporate, programme management or the customer can evaluate the project – and answer the key question about whether we've met our time, cost and quality objectives as discussed above – is via an **end project report** (see chapter 12). Among other things this contains the acceptance records and follow-on action recommendations already discussed, and also a formal **lessons report** derived from the entries in the lessons log. On a larger project it will be a composite of a number of individual and separate documents.

Lessons reports are now being prepared on a fairly widespread basis at the end of projects. The challenge remains as to how best to make sure they're effectively used. There is no simple answer to this, other than to suggest that thought be given to indexing entire reports based on the type and size of project, and the fields within them based on the type of lesson. Consideration could also be given to simply distributing any lessons report to all project managers in the host organisation at the end of each project, or even to having debrief meetings that all project managers can attend. The other problem is that a confident and proficient project manager might write an honest lessons report, only to have less confident members of the project board edit out most of the lessons concerning what could have been done better – which may significantly reduce the honesty and value of the report.

RECOMMEND PROJECT CLOSURE

The project manager will finally recommend to the project board that the project can be closed down, and if they agree their final decision represents the 'official close' of the project.

[The various outputs of closing a project are also shown in the completed Summary Diagram in Figure 19.]

18 CLOSING A PROJECT PROCESS

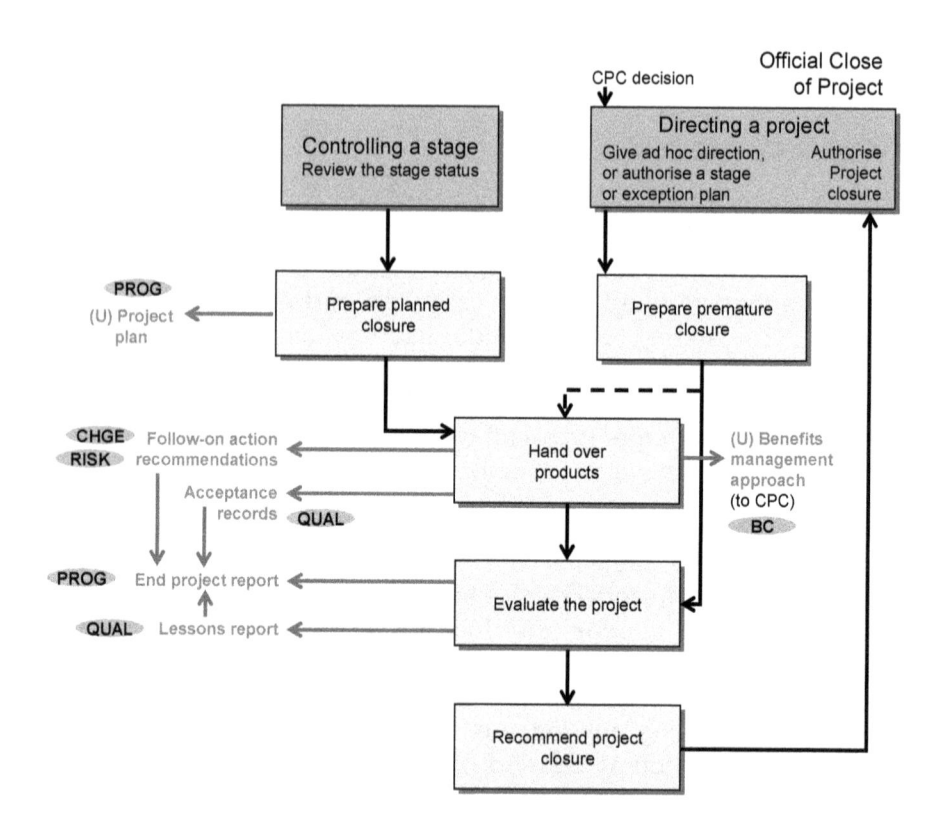

Figure 18: Closing a project process

(based on Figure 20.1 Managing Successful Projects with PRINCE2® 2017 Edition)

19 BRINGING IT ALL TOGETHER

Figure 19 represents the completed version of our unique Summary Diagram, which we've also been referring to at the end of each process chapter. It should act as a major aid to bringing together everything we've learned about PRINCE2. Remember the following explanations of its layout originally presented at the beginning of chapter 5:

- The four layers of the management structure are represented on the left-hand side.
- The top row of boxes represents the individual activities in the directing a project process, the middle row represents the processes used by the project manager, and the bottom row the process used by the team manager(s).
- Management products are shown next to the various arrows and are bolded. Most of these are created or updated by the project manager, as represented by anything next to an arrow that exits either above or below the middle row of process boxes. Those few created or updated by the team manager(s) are next to the right-hand arrow exiting upwards from the managing product delivery box on the bottom row.
- Management products next to arrows that don't 'go anywhere' or connect to another box are being created or updated in that process but not being given to anyone else.
- (U) means a management product is being 'updated'.
- The themes are shown in grey ovals. The abbreviations used are as per the list in chapter 2.
- If a management product is shown with a list of bullet points underneath it, this means it's a composite document that includes these others within it. On a larger project it will be a compilation of separate documents.
- Any writing in grey relates to 'project not as usual' or *exception* situations.

19 BRINGING IT ALL TOGETHER

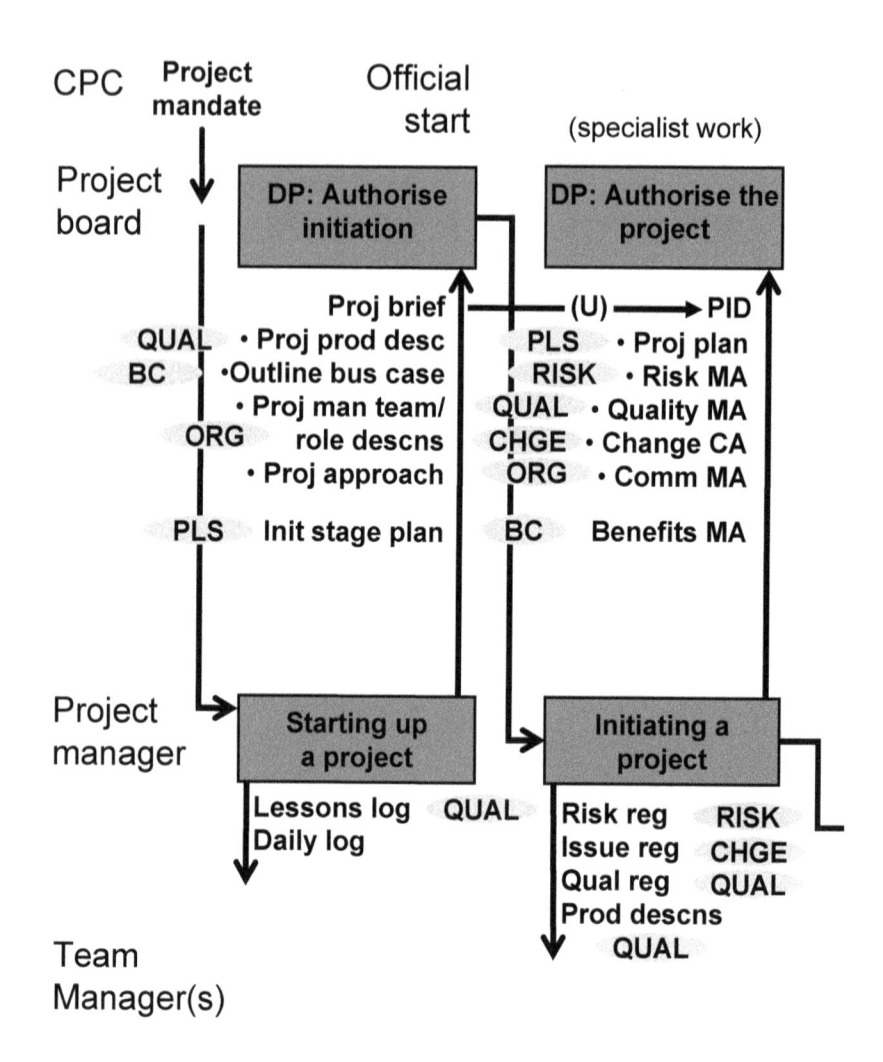

```
CPC    Project        Official
       mandate        start          (specialist work)

Project     ↓
board      [DP: Authorise         [DP: Authorise the
            initiation]            project]

              Proj brief        ─(U)──▶ PID
   QUAL  • Proj prod desc   PLS   • Proj plan
   BC    • Outline bus case RISK  • Risk MA
         • Proj man team/   QUAL  • Quality MA
   ORG     role descns      CHGE  • Change CA
         • Proj approach    ORG   • Comm MA

   PLS   Init stage plan    BC    Benefits MA


Project  ↘
manager    [Starting up          [Initiating a
            a project]            project]

           Lessons log  QUAL     Risk reg    RISK
           Daily log              Issue reg  CHGE
                                  Qual reg   QUAL
                                  Prod descns
Team                                         QUAL
Manager(s)
```

Figure 19: PRINCE2 Summary Diagram

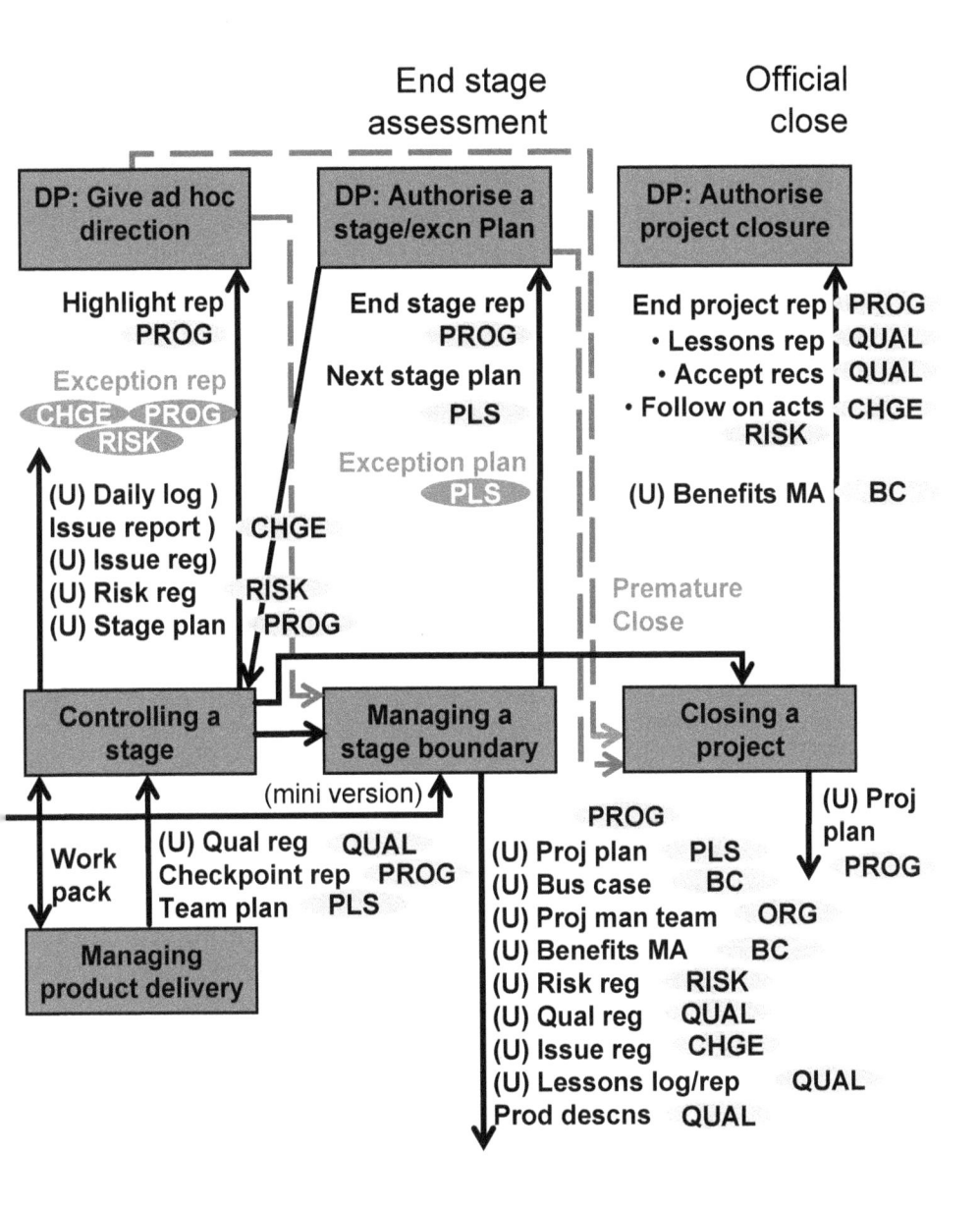

Figure 19: PRINCE2 Summary Diagram (cont)

20 TAILORING PRINCE2

We have already seen that if we <u>don't</u> tailor PRINCE2 it can be perceived as over-bureaucratic, too insistent on weighty documentation, and a 'sledgehammer to crack a nut'. Remember also that the seven principles of good project management practice are the only element that <u>can't</u> be tailored – because it's primarily adhering to these that allows us to say we're following PRINCE2.

For the rest of this chapter we'll look at some of the most important and commonplace ways in which we can tailor PRINCE2. A number of others are however discussed in chapters 4 and 21 of the official manual, which deal with tailoring PRINCE2 to the individual project and to the organisation as a whole, and embedding it therein. Further details can also be found in section 3 of each theme chapter and section 5 of each process chapter.

COMMON SENSE

As a general rule we should <u>never</u> be doing something when supposedly following PRINCE2 that doesn't make sense to us, or where we don't understand why we're doing it. <u>If</u> this happens it almost certainly means either we ourselves don't understand how to tailor PRINCE2, or our organisation doesn't and is forcing us to do things that are unnecessary.

NAMING CONVENTIONS

For the real world PRINCE2 doesn't care what you call things. It is only for exam purposes that you have to follow standard terminology, obviously. There is a downside to not using it in situations where, for example, you're working with external suppliers, or a project manager leaves or falls ill and you have to use an external replacement. But this can easily be mitigated by producing a glossary of terms that maps against PRINCE2. Having said that, we should at least be aiming for consistency within the host or supplier organisation.

MANAGEMENT PRODUCTS

Appendix A of the official manual provides details of the suggested *composition headings* for each of these. However we should be clear that the pertinent question is, "How formally do I need to write <u>this</u> document on <u>this</u> project?" We may decide we don't need to produce it formally at all on a simple project, or we may indeed decide to write it in full using all the standard headings on a larger, more complex project. But on many projects we will be somewhere in-between – in other words, we don't have to use all the headings all the time, only those that are relevant and appropriate in any given situation. As always common sense ought to be the key to how we tailor our documents. Plus, of course, we can actually call the headings by different names, as well as the documents themselves, as long as they contain the appropriate information to satisfy the principles of best practice. Moreover, we can of course combine management products where appropriate, or split them into separate documents.

Proformas for the management products with all the standard headings are available on the AXELOS website. However note that where proformas are used, especially as part of an in-house project management system, it's important that not all fields or headings have to be filled out formally – otherwise the system and therefore method will gain a reputation for overkill.

Finally we tend to refer to management products as documents, which suggests hard copy, but they can be presented in whatever way is appropriate. Alternatives include presentations slides, wall charts, and information held purely electronically in various project-related databases and systems.

PROCESSES AND ACTIVITIES

Any activity in a process can be split into separate more detailed activities if that makes sense in the organisational and project context. By contrast on smaller projects activities can be combined.

SIMPLE PROJECTS

The most obvious factor that will <u>always</u> play a part in determining

20 TAILORING PRINCE2

how formally we use the different elements of PRINCE2 will be the size or scale of the project. A simple project can be defined as one that is short in terms of timescale, low cost in terms of budget and low risk in terms of complexity. This will be relative to the size of the host organisation, but everyone will be able to form some judgement of what constitutes a simple project in their particular environment.

There are 4 main ways in which we tailor PRINCE2 for such a project:

- **Organisation theme** We can have a minimum of 2 people fulfilling all the roles: the executive taking on all the other project board roles and doing all their own assurance; and the project manager taking on the team manager's responsibilities, i.e. working directly with just one or two team members, and doing their own support (refer back to the roles above and below the 'line of independence' in Figure 6).
- **Starting up a project process** Because initiation itself will be short and low cost we can merge it with start up, and move straight from the mandate to a full PID.
- **Two stages** On a simple project stage 1 will be initiation, as always, but then there will just be one subsequent delivery stage. Accordingly end stage reports won't be required, nor will separate stage plans or work packages, because the project plan should suffice.
- **Management products** Given the above we can run a simple project using just 4 pieces of documentation:
    - A **PID**, which will contain as a minimum some sort of business case (even if it only gives the reason for doing the project and the cost and time budgets allocated to it); some sort of high level project plan (even if it's just a simple spreadsheet-based schedule); some sort of project product description providing some acceptance criteria (and possibly some individual product descriptions if there's more than one component product); and maybe something simple in respect of identification of stakeholders, benefits management and version control. What is

more, this time the PID will definitely be just a single document.
- **Highlight reports**, to allow the executive to monitor our progress during the single delivery stage (checkpoints with team members will be informal/verbal).
- A **daily log**, in which we'll record all issues, risks, lessons etc..
- An **end project report**, to close the project off properly, probably including some simple written acceptance of the project product.

There we have it. Just <u>two</u> people and <u>four</u> pieces of documentation to run a simple project. So don't ever let anyone tell you that PRINCE2 is an overly bureaucratic method.

It is probably also worth considering how some of the aspects of project management, which only tend to be undertaken with full formality on larger projects, are tailored for smaller ones. Project assurance, for example, can be carried out by the project board members themselves, and all they need to do is find some way of satisfying themselves that they can have confidence in their aspect of the project. The detailed approaches don't have to be formally and separately documented – we can simply make reference to corporate standards in each area, or in a smaller company have a brief discussion between members of the project management team to ensure we agree our approach to risk, quality, change and communication. Finally configuration management doesn't have to be fully formal on all projects, but we should at least make some arrangements for simple version control of products, and make sure we identify any linkages between them.

AGILE ENVIRONMENTS

Agile is an increasingly popular approach used for product development. In short it involves an iterative approach that uses 'sprints' or 'timeboxes', usually of between a week and a month in duration, to both design and build product(s) that must 'add value' (or benefit) in some way. To assist this process daily 'scrums' or

20 TAILORING PRINCE2

'standups' are held, which include the main user or 'product owner', who is regarded as an integral part of the team. This avoids the problems inherent in traditional 'waterfall-style' approaches, whereby after user requirements have been gathered the team goes away for a lengthy period to develop the required products, only to then find on presentation that they've misunderstood some, while others have changed in the interim.

Some people have tended to view agile as a 'fast on its feet' replacement for 'slow and ponderous' PRINCE2. This is undoubtedly a mistake, because agile is not a project management methodology – indeed nothing in agile makes mention of business cases, for example. Many organisations have already learned to their cost that on a project of any size and complexity the use of agile alone, without placing it under the umbrella of a proper project management methodology, can lead to a disastrous lack of control and direction. For this reason agile and PRINCE2 should be seen as complementary to each other. Agile can be used within the controlling a stage and especially managing product delivery processes for iterative, responsive design and development of specialist products. But without the proper use of other surrounding processes – start up, initiation, managing a stage boundary, closing a project and directing a project – the project may well fall short of its aims.

Aware that this combination will in many environments represent the future of best project practice, in 2016 AXELOS released a 'PRINCE2 Agile' guide and qualification. Among other things this tackles some of the harder aspects of integration. For example, agile teams are self-managing and responsibilities are shared – the 'scrum leader' is merely a facilitator. So careful thought needs to be given to how we incorporate the concept of project and team management into an agile environment, and how these roles might be involved in daily scrums. Having said that, PRINCE2's principle of managing by exception is ideally suited to an agile environment, where considerable authority is delegated to teams to get on with the job.

It is also worth mentioning that under an agile approach the time and cost of each sprint is fixed. This is because we never add new people into an agile team, certainly during a sprint, because this would disrupt its dynamics. But nor do we expect them to work overtime because they're apparently behind schedule. Instead we flex the quality and scope of what's being delivered in each sprint. This in turn requires the users to be ready to prioritise their requirements or 'user stories' into must-haves and merely nice-to-haves, and selections of both are chosen for each sprint to ensure flexibility. Remember too that within reason users may be changing their mind about what they want during a sprint, so flexibility on both sides is key.

For this reason change control is somewhat different in an agile environment, because regular change is a fully accepted part of the approach. Nevertheless care needs to be taken that change is still controlled to an appropriate degree, and especially that any significant changes requested by one particular product owner dovetail with the broader interests and requirements of other stakeholders and of the project as a whole.

For this reason too progress reporting will concentrate far more on the quality of products delivered and their functionality, rather than on the more conventional monitoring of time and cost. Moreover progress will tend to be displayed on large wall charts – such as 'burn charts', 'Kanban boards' or other 'information radiators' – allowing stakeholders to proactively 'pull' the information from them if they choose to.

In an agile environment the length of a sprint is usually too short, at a maximum of one month, for it to act effectively as a stage, so some other rationale for choosing stages has to be used. One possibility is if 2 or 3 sprints are combined into a 'release', which could then act as an obvious stage. However shorter stages may be required on high risk, experimental projects where the aim is to 'fail fast' if necessary.

In terms of closing a project down, much of the handover and acceptance of products has already been done. Moreover the use of 'retrospectives' at the end of each sprint automatically builds learning by experience into the project itself.

20 TAILORING PRINCE2

TAILORING TO AND EMBEDDING IN THE ORGANISATION

If an organisation truly desires to use PRINCE2 as its method for managing projects, providing consistency of approach across all departments, divisions and so on, then careful consideration must be given to the particular external and internal environment it operates in – in terms of processes, methods, standards and practices, regulatory requirements, and so on – and to how that influences its requirements of a project management method. PRINCE2 should then be tailored for the organisation as a whole to enable it to meet these requirements.

But producing a standardised approach to project management based on PRINCE2 isn't enough. Time and effort needs to be put into making sure the appropriate systems, documentation and so on are easily accessible, and that sufficient training is provided for appropriate staff, in all parts of the organisation. There has to be high level commitment of resources, led from the top. The organisation-wide implementation of the method can be handled as a project or programme in its own right, with stakeholder engagement and learning from experience key factors in promoting widespread use and continuous improvement.

Appendix 1 provides a questionnaire to help with tailoring PRINCE2 to any given project.

APPENDIX 1
PRINCE2 TAILORING QUESTIONNAIRE

WHAT KIND OF MANDATE HAVE WE BEEN GIVEN?

- Is it just a verbal and informal instruction? Is it more formal and extensive (for example if our project is part of a programme)? This will partly determine how much work we have to do in start-up and initiation, and even whether we can combine the two.

IS THIS A SMALL, MEDIUM OR LARGE PROJECT?

These figures are purely examples – each organisation to set its own limits:

- *Small projects* defined as e.g. costing from £0-10k and lasting from 1-2 months.
- *Medium projects* defined as e.g. costing from £10-100k and lasting from 2-6 months.
- *Large projects* defined as e.g. costing over £100k and lasting more than 6 months.

HOW MANY PEOPLE WILL MAKE UP THE PROJECT TEAM?

- Is it sufficient for us to just have an Executive who takes on all the Board roles?
- If not can the Executive at least take on the Senior User role while we have one or more separate Senior Suppliers?
- Can the Board members undertake their own Assurance?
- Do we need separate Team Managers?
- Can the Project Manager undertake their own Project Support?
- Can the Project Board handle all changes, or can the Project Manager act as the Change Authority?

HOW MANY STAGES DO WE NEED?

- On a small project can we combine start up and initiation? Or for any other reason can we combine them (e.g. because this is a compulsory/legislative project)?

APPENDIX 1: PRINCE2 TAILORING QUESTIONNAIRE

- On a small project can we just use two stages, for initiation and then the main delivery stage? If so we won't need to use stage plans or the stage boundary process.
- Are we adopting a waterfall or agile approach? If agile, will start-up be a stage in its own right in which we produce the vision and product roadmap, while in initiation we produce the product backlog? Or will it again be better to combine the two? Will stage boundaries coincide with the end of every sprint, or every release, or some other factor?
- Are there important decision, risk or annual budget points that need a stage boundary?

WHAT DOCUMENTATION DO WE NEED?

Reference Appendix A of *Managing Successful Projects With PRINCE2*.

- **For a small project**
    - A single-document **PID** which will include a simple **business case**, a simple **project plan**, a simple **project product description**, possibly a few individual simplified **product descriptions**, and maybe something simple in respect of identification of stakeholders, benefits management and version control.
    - A **daily log** that will include all issues, risks, lessons etc..
    - Cut-down **highlight reports** for the main delivery stage.
    - A cut-down **end project report**.
- **For a medium project**

    Baselined documents subject to change control
    - A **project brief** and a **PID** containing various of the following:
    - Reasonably comprehensive **project product description**, **business case**, **project/stage plans** and individual **product descriptions**.
    - Cut-down **communication, benefits, risk, quality** and **change management approaches** and **work packages**.

Records
- ○ Comprehensive **issue/risk registers** and **lessons log**.
- ○ Optional 'official' **daily log**, **quality register** and **configuration item records**.

Reports
- ○ Tailored **checkpoint, highlight, exception, end stage** and **end project reports**.
- ○ Optional **issue/lessons reports** and **product status accounts**.

➤ **For a large project**
- ○ The full range of documentation included in Appendix A is likely to be required, using most of the headings in each although still tailored to suit.

HOW WILL WE PRESENT OUR DOCUMENTS?

➤ Do they all have to be formal printed documents, or can some be slide presentations, wall charts or held in databases or other IT systems?

➤ Can we usefully use agile-type displays such as Burn charts and Kanban boards, and let stakeholders pull progress information from them rather than have it pushed at them in formal reports?

WHAT TOLERANCES WILL WE USE?

➤ *Time and cost tolerances* should be pretty much mandatory on all projects of whatever size if we're to follow the principle of managing by exception. Remember that we can proactively set plus *zero* tolerances if a project absolutely must come in on time or budget. As for minus tolerances, we don't have to use these, but we will if we want to be informed that something is going to come in especially early or under budget – if, for example, we would then need to reallocate resources or unused budgets.

➤ *Other tolerance areas* should be used as appropriate, but will come more into play the larger and more complex the project.

APPENDIX 2
PRINCE2 PRINCIPLES CHECKLIST

The following is a checklist of questions related to whether or not an organisation is following the 7 principles of best practice – which, remember, are the main determinant of whether or not projects are being managed using PRINCE2.

CONTINUED BUSINESS JUSTIFICATION: Are business cases being properly prepared? Are they being updated, including benefits forecasts? How often? Are they ever used to shut a project down early? Are benefits reviews being held?

MANAGE BY STAGES: Are projects being split into stages? Are separate/more detailed stage plans being created? Are there clear management stage boundaries? Are assessments of ongoing viability held at these boundaries based on an updated business case?

MANAGE BY EXCEPTION: Are measurable cost and time tolerances being set at different levels, or is just RAG reporting being used? If the latter, who decides what's red or amber and is it just based on discretion? Are regular meetings being held with the project board even within a stage, and if so are any decisions being made or are they just progress meetings? How many people are on the project board, and is it more of a talking shop than a decision-making body?

DEFINED ROLES AND RESPONSIBILITIES: Are the project board and corporate, programme management or the customer properly differentiated? Is independent project assurance being used? Is there only ever one genuine project manager on any one project, or are multiple project managers trying to share the responsibilities alongside each other? Are separate team managers used where appropriate?

FOCUS ON PRODUCTS: Are product descriptions being properly prepared, including measurable acceptance/quality criteria and tolerances? Is product-based planning being used, especially workshops to create product breakdown structures? If so are both users and suppliers involved?

LEARN FROM EXPERIENCE: Are lessons logs being kept and lessons reports being produced? How honest are the latter? How well are they used? Are lessons from previous projects being captured up front?

TAILOR TO THE PROJECT: Are the formality and contents of management products being varied? Is terminology being adjusted to suit the organisation? Is the method being used flexibly on simpler projects? Are agile and PRINCE2 being effectively combined? Is there any attempt to properly embed a tailored version of PRINCE2 into the entire organisation's ways of working?

INDEX

acceptance criteria, 3, 15, 23, 36, 59, 90
acceptance records, 38, 63, 82-3
agile, 10, 42, 66, 91-4, 98
APMG, 1
AXELOS, 1, 89, 92

benefits management approach, 29-30, 37, 69, 78, 83, 90, 98
business case, 3-9, 12, 15, 19-20, 23-6, 28-30, 36, 41, 54-6, 58-61, 64-6, 69-70, 78-81, 90, 92
business case theme, 12

change authority, 32, 55
change budget, 50, 55
change control, 37, 52-3, 55-7, 76, 93
change control approach, 56
change theme, 14, 52, 76
checkpoint report, 24-5, 60-1, 72, 74
closing a project process, 9, 17, 20, 25, 38, 62, 76, 81-3, 92-3
communication management approach, 34, 37
configuration item record, 56, 69, 78
configuration management, 32, 37, 52, 55-7, 68, 72, 91
continued business justification principle, 4, 8, 15, 20, 25-6, 41, 69, 79
controlling a stage process, 9, 16, 24, 40, 55, 62, 71-2, 74, 76, 79, 92
corporate, programme management or the customer, 13, 23, 26, 30-1, 34, 38, 41, 59-61, 64, 66, 79-83

daily log, 54, 62, 65-6, 74, 91
delivery step, 42
directing a project process, 9, 17, 22-3, 61, 64, 66, 85, 92

end project report, 25, 50, 83, 91
end stage assessment, 20, 25, 69
end stage report, 25, 60, 69, 79, 90
event-driven control, 58, 62
exception report, 25, 55, 60, 62, 76, 80
executive, 13, 23, 31, 33, 64, 65-6, 90-1
external products, 44

focus on products principle, 43
follow-on action recommendations, 82-3

highlight report, 24-5, 32, 34, 38, 60-1, 75

initiating a project process, 9,

16, 19, 23, 26, 28-9, 36, 37, 41, 44, 50, 64-6, 68-70, 78, 90, 92
integrated elements, 8-9, 18
issue register, 53-4, 68, 74, 79, 82
issue report, 54, 74-5

learn from experience principle, 29
lessons log, 65, 79, 83, 96, 99
lessons report, 63, 65, 79, 83

manage by exception principle, 8, 14-15, 23, 25-6, 34, 40-1, 55, 59-62, 74-5, 80, 82, 85, 92
management product, 9, 14-15, 19, 22-4, 28-9, 31, 37, 56, 65, 85, 89
management stage, 42, 95
managing a stage boundary process, 9, 16, 20, 25, 62, 69, 76, 78, 80, 92
managing product delivery process, 9, 16, 22, 38, 40, 62, 71-3, 85, 92

off-specification, 54-5
opportunity, 47, 49
organisation theme, 12

plans theme, 15-16
premature close, 6, 20, 26, 61, 79, 82
product breakdown structure, 44, 68, 78
product description, 36, 43-4, 53, 56, 58, 68, 72, 78, 90

product flow diagram, 44
product-based planning technique, 16, 36, 43-4, 46, 68, 78
product status account, 99
programme board, 13, 59
progress report, 9, 14, 24-5, 34, 50, 62, 93
progress theme, 15, 58
project approach, 66, 69
project assurance, 32-3, 38, 55
project board, 13-15, 17, 19, 20, 23-6, 31-4, 38, 40-1, 54-5, 59-61, 64-6, 68, 70, 74-5, 79-80, 82-3, 90-1
project brief, 19, 23, 59, 65-6, 69, 91
project controls, 68-9
project initiation documentation, 19, 23, 65, 68-9, 79, 90
project issue, 14, 24, 52
project management team, 25, 30, 32, 38, 65-6, 69, 78, 81, 91
project manager, 1-3, 5, 13-14, 16, 19-20, 22-4, 31-4, 36-7, 40, 43, 52, 54-5, 59-60, 62, 65-6, 72-3, 75, 78, 81-3, 85, 88, 90
project mandate, 4, 19, 23, 64-5
project plan, 15-16, 20, 24-6, 40-1, 43-5, 56, 58, 61, 64, 68-9, 78-82, 90
project product, 10, 15, 23, 28, 36, 38, 43-4, 56, 58-9, 64-6, 69, 81-2, 90-1

INDEX

project product description, 23, 36, 43, 56, 58-9, 64-6, 69, 90
project support, 30, 32, 38

quality assurance, 38
quality control, 37, 72
quality management approach, 37
quality management system, 38
quality planning, 37
quality register, 37, 73, 78
quality theme, 16, 37, 43
quality-cost-time triangle, 4-5, 23, 44, 64

RAG reporting, 14
request for change, 52-5, 82
risk actionee, 50
risk appetite, 47
risk budget, 50, 55
risk management approach, 50, 57
risk owner, 49
risk register, 50, 53, 66, 74, 79, 82
risk theme, 12, 25, 66
role descriptions, 65-6, 69

scope, 4-6, 26, 43-5, 52, 55, 61, 79, 93
scope creep, 52, 55
senior supplier, 13, 31, 78
senior user, 13, 29, 31, 81
specialist product, 9-10, 13-16, 20, 31, 36-7, 43-5, 52-3, 56, 68, 72, 78, 92
stage boundary, 15, 20, 24, 26, 28, 33, 37, 40-2, 44, 70, 79-80
stage plan, 15, 24-6, 40, 45, 54, 58-60, 66, 69-70, 74, 78-9, 90
stakeholder, 30, 34, 37, 44, 52, 81, 93-4
starting up a project process, 9, 16, 19-20, 23-4, 28, 36, 60, 64-6, 68-9, 90, 92

tailor to the project principle, 1, 9-10, 19, 32, 53, 62, 64, 68-9, 88-91, 94
team manager, 16, 20, 22, 24, 31-3, 37, 40, 44, 53, 60, 62, 71-5, 78, 85, 90
team member, 14, 31, 33, 52-3, 62, 72, 90-1
team members, 14, 31, 33, 52, 62, 72, 90-1
team plan, 40-1, 44, 58, 60, 72
threat, 41, 47-9
time-driven control, 61
tolerance, 14-15, 25-6, 28, 37, 40-3, 53-5, 58-63, 72, 75, 80

work package, 24-5, 37, 40-2, 53-4, 58-62, 71-5, 90

IAN LAWTON was born in 1959. Formerly a chartered accountant, sales executive, business consultant and IT project manager, since 2008 he has trained more than 2000 delegates on PRINCE2 courses. He has worked for some of the market leaders in training provision such as Parity and QA, and understands how to make the method accessible to newcomers.

www.prince2madesimple.co.uk

Lightning Source UK Ltd.
Milton Keynes UK
UKHW020645230321
380840UK00009B/409